Stories of the L

Frank Mundell

Alpha Editions

This edition published in 2024

ISBN : 9789362920959

Design and Setting By
Alpha Editions
www.alphaedis.com
Email - info@alphaedis.com

Contents

PREFACE

In sending forth this little work to the public, I desire to acknowledge my obligations to the following:--The Royal National Lifeboat Institution for the valuable matter placed at my disposal, also for the use of the illustrations on pages 20 and 21; to Mr. Clement Scott and the proprietors of *Punch* for permission to use the poem, "The Warriors of the Sea"; to the proprietors of *The Star* for the poem, "The Stranding of the *Eider*"; and to the proprietors of the *Kent Argus* for so freely granting access to the files of their journal. Lastly, my thanks are due to the publishers--at whose suggestion the work was undertaken--for the generous manner in which they have illustrated the book.

F. M.
LONDON, *September,* 1894.

CHAPTER I.
MAN THE LIFEBOAT!

Go Lionel Lukin, a coachbuilder of Long Acre, London, belongs the honour of inventing the lifeboat. As early as the year 1784 he designed and fitted a boat, which was intended "to save the lives of mariners wrecked on the coast." It had a projecting gunwale of cork, and air-tight lockers or enclosures under the seats. These gave the boat great buoyancy, but it was liable to be disabled by having the sides stove in. Though Lukin was encouraged in his efforts by the Prince of Wales--afterwards George the Fourth--his invention did not meet with the approval of those in power at the Admiralty, and Lukin's only lifeboat which came into use was a coble that he fitted up for the Rev. Dr. Shairp of Bamborough. For many years this was the only lifeboat on the coast, and it is said to have saved many lives.

In the churchyard of Hythe, in Kent, the following inscription may be read on the tombstone, which marks the last resting-place of the "Father of the Lifeboat":--

"This LIONEL LUKIN
was the first who built a lifeboat, and was the
original inventor of that quality of safety, by
which many lives and much property have been
preserved from shipwreck, and he obtained for
it the King's Patent in the year 1785."

The honour of having been the first inventor of the lifeboat is also claimed by two other men. In the parish church of St. Hilda, South Shields, there is a stone "Sacred to the Memory of William Wouldhave, who died September 28, 1821, aged 70 years, Clerk of this Church, and Inventor of that invaluable blessing to mankind, the Lifeboat." Another similar record tells us that "Mr. Henry Greathead, a shrewd boatbuilder at South Shields, has very generally been credited with designing and building the first lifeboat, about the year 1789." As we have seen, Lukin had received the king's patent for his invention four years before Greathead brought forward his plan. This proves conclusively that the proud distinction belongs by right to Lionel Lukin.

In September 1789 a terrible wreck took place at the mouth of the Tyne. The ship *Adventure* of Newcastle went aground on the Herd Sands, within three hundred yards of the shore. The crew took to the rigging, where they remained till, benumbed by cold and exhaustion, they dropped one by one into the midst of the tremendous breakers, and were drowned in the presence of thousands of spectators, who were powerless to render them any assistance.

Deeply impressed by this melancholy catastrophe, the gentlemen of South Shields called a meeting, and offered prizes for the best model of a lifeboat "calculated to brave the dangers of the sea, particularly of broken water." From the many plans sent in, those of William Wouldhave and Henry Greathead were selected, and after due consideration the prize was awarded to "the shrewd boatbuilder at South Shields." He was instructed to build a boat on his own plan with several of Wouldhave's ideas introduced. This boat had five thwarts, or seats for rowers, double banked, to be manned by ten oars. It was lined with cork, and had a cork fender or pad outside, 16 inches deep. The chief point about Greathead's invention was that the keel was curved instead of being straight. This circumstance, simple as it appears, caused him to be regarded as the inventor of the first practicable lifeboat, for experience has proved that a boat with a curved keel is much more easily launched and beached than one with a straight keel.

Lifeboats on this plan were afterwards placed on different parts of the coast, and were the means of saving altogether some hundreds of lives. By the end of the year 1803 Greathead had built no fewer than thirty-one lifeboats, eight of which were sent to foreign countries. He applied to Parliament for a national reward, and received the sum of £1200. The Trinity House and Lloyd's each gave him £105. From the Society of Arts he received a gold medal and fifty guineas, and a diamond ring from the Emperor of Russia.

The attention thus drawn to the needs of the shipwrecked mariner might have been expected to be productive of good results, but, unfortunately, it was not so. The chief reason for this apathy is probably to be found in the fact that, though the lifeboats had done much good work, several serious disasters had befallen them, which caused many people to regard the remedy as worse than the disease. Of this there was a deplorable instance in 1810, when one of Greathead's lifeboats, manned by fifteen men, went out to the rescue of some fishermen who had been caught in a gale off Tynemouth. They succeeded in taking the men on board, but on nearing the shore a huge wave swept the lifeboat on to a reef of rocks, where it was smashed to atoms. Thirty-four poor fellows--the rescued and the rescuers-- were drowned.

It was not until twelve years after this that the subject of the preservation of life from shipwreck on our coast was successfully taken up. Sir William Hillary, himself a lifeboat hero, published a striking appeal to the nation on behalf of the perishing mariner, and as the result of his exertions the Royal

National Institution for the Preservation of Life from Shipwreck was established in 1824. This Society still exists under the well-known name of the Royal National Lifeboat Institution. It commenced its splendid career with about £10,000, and in its first year built and stationed a dozen lifeboats on different parts of the coast.

For many years the Society did good work, though sadly crippled for want of funds. In 1850 the Duke of Northumberland offered the sum of one hundred guineas for the best model of a lifeboat. Not only from all parts of Great Britain, but also from America, France, Holland, and Germany, plans and models were sent in to the number of two hundred and eighty. After six months' examination, the prize was awarded to James Beeching of Great Yarmouth, and his was the first self-righting lifeboat ever built. The committee were not altogether satisfied with Beeching's boat, and Mr. Peake, of Her Majesty's Dockyard at Woolwich, was instructed to design a boat embodying all the best features in the plans which had been sent in. This was accordingly done, and his model, gradually improved as time went on, was adopted by the Institution for their boats.

LAUNCHING THE LIFEBOAT

LAUNCHING THE LIFEBOAT

The lifeboats now in use measure from 30 to 40 feet in length, and 8 in breadth. Buoyancy is obtained by air-chambers at the ends and on both sides. The two large air-chambers at the stem and stern, together with a heavy iron keel, make the boat self-righting, so that should she be upset she cannot remain bottom up. Between the floor and the outer skin of the boat there is a space stuffed with cork and light hard wood, so that even if a hole was made in the outer covering the boat would not sink. To insure the safety of the crew in the event of a sea being shipped, the floor is pierced with holes, into which are placed tubes communicating with the sea, and valves so arranged that the water cannot come up into the boat, but should she ship a sea the valves open downwards and drain off the water. A new departure in lifeboat construction was made in 1890, when a steam lifeboat, named the Duke of Northumberland, was launched. Since then it has saved many lives, and has proved itself to be a thoroughly good sea boat. While an ordinary lifeboat is obliged to beat about and lose valuable time, the steam lifeboat goes straight to its mark even in the roughest sea, so that probably before long the use of steam in combating the storm will become general.

Nearly every lifeboat is provided with a transporting carriage on which she constantly stands ready to be launched at a moment's notice. By means of this carriage, which is simply a framework on four wheels, the lifeboat can be used along a greater extent of coast than would otherwise be possible. It is quicker and less laborious to convey the boat by land to the point nearest the wreck, than to proceed by sea, perhaps in the teeth of a furious gale. In addition to this a carriage is of great use in launching a boat from the beach, and there are instances on record when, but for the carriage, it would have been impossible for the lifeboat to leave the shore on account of the high surf.

THE LIFEBOAT HOUSE.

THE LIFEBOAT HOUSE.

The boats belonging to the National Lifeboat Institution are kept in roomy and substantial boathouses under lock and key. The coxswain has full charge of the boat, both when afloat and ashore. He receives a salary of £8 a year, and his assistant £2 a year. The crew of the lifeboat consists of a bowman and as many men as the boat pulls oars. On every occasion of going afloat to save life, each man receives ten shillings, if by day; and £1, if by night. This money is paid to the men out of the funds of the Institution, whether they have been successful or not. During the winter months these payments are now increased by one half.

MEDAL OF THE ROYAL NATIONAL LIFEBOAT INSTITUTION.

MEDAL OF THE ROYAL NATIONAL LIFEBOAT INSTITUTION.

The cost of a boat with its equipment of stores--cork lifebelts, anchors, lines, lifebuoys, lanterns, and other articles--is upwards of £700, and the expense of building the boathouse amounts to £300, while the cost of maintaining it is £70 a year. The Institution also awards medals to those who have distinguished themselves by their bravery in saving life from shipwreck. One side of this medal is adorned with a bust of Her Majesty, Queen Victoria, who is the patroness of the Institution. The other side represents three sailors in a lifeboat, one of whom is rescuing an exhausted mariner from the waves with the inscription, "Let not the deep swallow me up." Additional displays of heroism are rewarded by clasps bearing the number of the service.

"When we think of the vast extent of our dangerous coasts, and of our immense interest in shipping, averaging arrivals and departures of some 600,000 vessels a year; when we think of the number of lives engaged, some 200,000 men and boys, besides untold thousands of passengers, and goods amounting to many millions of pounds in value, the immense importance of the lifeboat service cannot be over-estimated." Well may we then, "when the storm howls loudest," pray that God will bless that noble Society, and the band of humble heroes who man the three hundred lifeboats stationed around the coasts of the British Isles.

CHAPTER II.
LIFEBOAT DISASTERS.

We have already referred to the numerous disasters which did so much to retard the progress of the lifeboat movement. Now let us see how these disasters were caused. The early lifeboats, though provided with a great amount of buoyancy, had no means of freeing themselves of water, or of self-righting if upset, and the absence of these qualities caused the loss of many lives.

Sir William Hillary, who may be regarded as the founder of the National Lifeboat Institution, distinguished himself, while living on the Isle of Man, by his bravery in rescuing shipwrecked crews. It was estimated that in twenty-five years upwards of a hundred and forty vessels were wrecked on the island, and a hundred and seventy lives were lost; while the destruction of property was put down at a quarter of a million. In 1825, when the steamer *City of Glasgow* went ashore in Douglas Bay, Sir William Hillary went out in the lifeboat and assisted in taking sixty-two people off the wreck. In the same year the brig *Leopard* went ashore, and Sir William again went to the rescue and saved eleven lives. While he lived on the island, hardly a year passed without him adding fresh laurels to his name, and never did knight of old rush into the fray with greater ardour than did this gallant knight of the nineteenth century to the rescue of those in peril on the sea. His greatest triumph, however, was on the 20th of November 1830, when the mail steamer *St. George* stranded on St. Mary's Rock and became a total wreck. The whole crew, twenty-two in number, were rescued by the lifeboat. On this occasion he was washed overboard among the wreck, and it was with the greatest difficulty that he was saved, having had six of his ribs broken.

In 1843 the lifeboat stationed at Robin Hood Bay went out to the assistance of the *Ann* of London. Without mishap the wreck was reached, and the work of rescue was begun. Several of the shipwrecked men jumped into the boat just as a great wave struck her, and she upset. Some of the crew managed to scramble on to the bottom of the upturned boat and clung to the keel for their lives.

The accident had been witnessed by the men on the beach, and five of them immediately put out to the rescue. They had hardly left the shore when an enormous sea swept down upon them, causing the boat to turn a double somersault, and drowning two of the crew. Altogether twelve men lost their lives on this occasion. Those who were saved floated ashore on the bottom of the lifeboat.

The Herd Sand, memorable as the scene of the wreck of the *Adventure*, witnessed a lamentable disaster in 1849, when the *Betsy* of Littlehampton went aground. The South Shields lifeboat, manned by twenty-four experienced pilots, went out to the rescue. While preparing to take the crew on board, she was struck by a heavy sea, and before she could recover herself, a second mighty wave threw her over. Twenty out of the twenty-four of her crew were drowned. The remainder and the crew of the *Betsy* were rescued by two other lifeboats, which put off from the shore immediately upon witnessing what had happened.

The advantages of the self-righting and self-emptying boats may be best judged from the fact, that since their introduction in 1852, as many as seventy thousand men have gone out in these boats on service, and of these only seventy-nine have nobly perished in their gallant attempts to rescue others. This is equal to a loss of one man in every eight hundred and eighty.

During the terrible storm which swept down upon our coast in 1864, the steamer *Stanley* of Aberdeen was wrecked while trying to enter the Tyne. The *Constance* lifeboat was launched from Tynemouth, and proceeded to the scene of the wreck. The night was as dark as pitch, and from the moment that the boat started, nothing was to be seen but the white flash of the sea, which broke over the boat and drenched the crew. As quickly as she freed herself of water, she was buried again and again. At length the wreck was reached, and while the men were waiting for a rope to be passed to them, a gigantic wave burst over the *Stanley* and buried the lifeboat. Every oar was snapped off at the gunwale, and the outer ends were swept away, leaving nothing but the handles. When the men made a grasp for the spare oars they only got two--the remainder had been washed overboard.

It was almost impossible to work the *Constance* with the rudder and two oars, and while she was in this disabled condition a second wave burst upon her. Four of the crew either jumped or were thrown out of the boat, and vanished from sight. A third mighty billow swept the lifeboat away from the wreck, and it was with the utmost difficulty that she was brought to land. Two of the men, who had been washed out of the boat, reached the shore in safety, having been kept afloat by their lifebelts. The other two were drowned.

Speaking of the attempted rescue, the coxswain of the *Constance* said: "Although this misfortune has befallen us, it has given fresh vigour to the crew of the lifeboat. Every man here is ready, should he be called on again, to act a similar part."

Thirty-five of those on board the *Stanley*, out of a total number of sixty persons, were afterwards saved by means of ropes from the shore.

One of the most heartrending disasters, which have befallen the modern lifeboat, happened on the night of the 9th of December 1886. The lifeboats at Southport and St. Anne's went out in a furious gale to rescue the crew of a German vessel named the *Mexico*. Both were capsized, and twenty-seven out of the twenty-nine who manned them were drowned. It was afterwards found out that the Southport boat succeeded in making the wreck, and was about to let down her anchor when she was capsized by a heavy sea. Contrary to all expectations the boat did not right, being probably prevented from doing so by the weight of the anchor which went overboard when the boat upset.

What happened to the St. Anne's lifeboat can never be known, for not one of her crew was saved to tell the tale. It is supposed that she met with some accident while crossing a sandbank, for, shortly after she had been launched, signals of distress were observed in that quarter. Next morning the boat was found on the beach bottom up with three of her crew hanging to the thwarts--dead.

NEWS OF A WRECK ON THE COAST.

Such is the fate that even to-day overhangs the lifeboatman on the uncertain sea. Yet he is ever ready on the first signal of distress to imperil his life to rescue the stranger and the foreigner from a watery grave. "First come, first in," is the rule, and to see the gallant lifeboatmen rushing at the top of their speed in the direction of the boathouse, one would imagine that they were hurrying to some grand entertainment instead of into the very jaws of death. It is not for money that they thus risk their lives, as the pay they receive is very small for the work they have to perform. They are indeed heroes, in the truest sense of the word, and give to the world a glorious example of duty well and nobly done.

CHAPTER III.
THE WARRIORS OF THE SEA.

[On the night of the 9th of December 1886, the Lytham, Southport, and St. Anne's lifeboats put out to rescue the crew of the ship *Mexico*, which had run aground off the coast of Lancashire. The Southport and St. Anne's boats were lost, but the Lytham boat effected the rescue in safety.]

Up goes the Lytham signal!

St. Anne's has summoned hands!

Knee deep in surf the lifeboat's launched

Abreast of Southport sands!

Half deafened by the screaming wind,

Half blinded by the rain,

Three crews await their coxswains,

And face the hurricane!

The stakes are death or duty!

No man has answered "No"!

Lives must be saved out yonder

On the doomed ship *Mexico*!

Did ever night look blacker?

Did sea so hiss before?

Did ever women's voices wail

More piteous on the shore?

Out from three ports of Lancashire

That night went lifeboats three,

To fight a splendid battle, manned

By "Warriors of the Sea."

Along the sands of Southport

Brave women held their breath,

For they knew that those who loved them

Were fighting hard with death;

A cheer went out from Lytham!

The tempest tossed it back,

As the gallant lads of Lancashire

Bent to the waves' attack;

And girls who dwelt about St. Anne's,

With faces white with fright,

Prayed God would still the tempest

That dark December night.

Sons, husbands, lovers, brothers,

They'd given up their all,

These noble English women

Heartsick at duty's call;

But not a cheer, or tear, or prayer,

From those who bent the knee,

Came out across the waves to nerve

Those Warriors of the Sea.

Three boats went out from Lancashire,

But one came back to tell

The story of that hurricane,

The tale of ocean's hell!

All safely reached the *Mexico*,

Their trysting-place to keep;

For one there was the rescue,

The others in the deep

Fell in the arms of victory

Dropped to their lonely grave,
Their passing bell the tempest,
Their requiem the wave!
They clung to life like sailors,
They fell to death like men,--
Where, in our roll of heroes,
When in our story, when,
Have Englishmen been braver,
Or fought more loyally
With death that comes by duty
To the Warriors of the Sea?

One boat came back to Lytham
Its noble duty done;
But at St. Anne's and Southport
The prize of death was won!
Won by those gallant fellows
Who went men's lives to save,
And died there crowned with glory,
Enthroned upon the wave!
Within a rope's throw off the wreck
The English sailors fell,
A blessing on their faithful lips,
When ocean rang their knell.
Weep not for them, dear women!
Cease wringing of your hands!
Go out to meet your heroes
Across the Southport sands!
Grim death for them is stingless!

The grave has victory!

Cross oars and bear them nobly home,

Brave Warriors of the Sea!

When in dark nights of winter

Fierce storms of wind and rain

Howl round the cosy homestead,

And lash the window-pane--

When over hill and tree top

We hear the tempests roar,

And hurricanes go sweeping on

From valley to the shore--

When nature seems to stand at bay,

And silent terror comes,

And those we love on earth the best

Are gathered in our homes,--

Think of the sailors round the coast,

Who, braving sleet or snow,

Leave sweethearts, wives, and little ones

When duty bids them go!

Think of our sea-girt island!

A harbour, where alone

No Englishman to save a life

Has failed to risk his own.

Then when the storm howls loudest,

Pray of your charity

That God will bless the lifeboat

And the Warriors of the Sea!

CLEMENT SCOTT.

(By permission of the Author, and the Proprietors of "Punch.")

CHAPTER IV.
THE GOODWIN SANDS.

About six miles off the east coast of Kent there is a sandbank known as the Goodwin Sands, extending for a distance of ten miles, between the North Foreland and the South Foreland. No part of our coast is so much dreaded by the mariner, and from early times it has been the scene of many terrible disasters. As Shakespeare says, it is "a very dangerous flat, and fatal, where the carcasses of many a tall ship lie buried."

It is said that the site of the Goodwin Sands was at one time occupied by a low fertile island, called Lomea, and here lived the famous Earl Godwin. After the Battle of Hastings, William the Conqueror took possession of these estates, and bestowed them, as was the custom in those days, upon the Abbey of St. Augustine at Canterbury. The abbot, however, seems to have had little regard for the property, and he used the funds with which it should have been maintained in building a steeple at Tenterden, an inland town near the south-west border of Kent. The wall, which defended the island from the sea, being thus allowed to fall into a state of decay, was unable to withstand the storm that, in 1099, burst over Northern Europe, and the waves rushed in and overwhelmed the island. This gave rise to the saying, "Tenterden steeple was the cause of the Goodwin Sands."

At high tide the whole of this dangerous shoal is covered by the sea to the depth of several feet; but at low water large stretches of sand are left hard and dry. At such a time it is perfectly safe for anyone to walk along this island desert for miles, and cricket is known to have been played in some places. Here and there the surface is broken by large hollows filled with water. Should the visitor, however, attempt to wade to the opposite side, he is glad to beat a hasty retreat, as he finds himself sinking with alarming rapidity into the sand, which the action of the water has rendered soft.

Between the Goodwins and the coast of Kent is the wide and secure roadstead called the Downs. Here, when easterly or south-easterly winds are blowing, ships may ride safely at anchor; but when a storm comes from the west, vessels are no longer secure, and frequently break from their moorings and become total wrecks on the sands. To warn mariners of their danger, four lightships are anchored on different parts of the sands. Each is provided with powerful lanterns, the light of which can be seen, in clear weather, ten miles off. During foggy weather, fog sirens are sounded and gongs are beaten to tell the sailor of his whereabouts. Notwithstanding all these precautions, the number of vessels stranded on the Goodwins every year is appalling; and but for the heroic efforts of the Kentish lifeboatmen, the loss of life would be still more terrible.

The work done by the boatmen all around our coast cannot be too highly estimated, but a special word of praise is due to the Ramsgate men. They have, without doubt, saved more lives than the men of any other port in the kingdom. Being stationed so near to the deadly Goodwins has given them greater opportunities for service, and they have also a steam tug in attendance on the lifeboat to tow her to the scene of disaster. So that, no matter what is the direction of the wind, they can always go out.

Recently, I went down to this "metropolis of the lifeboat service," for the express purpose of interviewing one of those warriors of the sea. The place was crowded with holiday-makers, and the harbour presented a busy scene. Four fine large yachts were getting their passengers on board for "a two-hours' sail." A yellow-painted tug was puffing to and fro, towing coasting vessels and luggers out of the harbour, and threatening to run down several small boats which repeatedly tried to cross her bows. At some distance from where I was standing lay the lifeboat *Bradford*, motionless and neglected, and looking strangely out of place in such smooth water. How the sight of the boat recalled to my mind all that I had ever read or heard of the perils of "those who go down to the sea in ships"--the storm, the wreck, the dark winter night, the midnight summons to man the lifeboat, the struggle for a place, the sufferings from cold, the happy return with the crew all saved,--these and other similar incidents seemed to pass before my eyes like a panorama--the centre object ever being the blue-painted *Bradford*.

"Have a boat this morning, sir?" said a thick muffled voice quite close to me. Turning round I saw a little, old man with a bronzed, weather-beaten face.

"Not this morning, thank you," I replied; "unless you will let me have the lifeboat for an hour or two."

He shook his head and turned away. Then it suddenly seemed to strike him that possibly I did not know the uses of the lifeboat, and would be none the worse if I received a little information on the subject.

A RAMSGATE BOATMAN

A RAMSGATE BOATMAN

"The lifeboat's not a pleasure boat, sir," he said, "and never goes out unless in cases of distress. I reckon if you went out in lifeboat weather once, you'd never want to go again."

"I suppose you have heavy seas here at times?" I remarked.

"Nobody that hasn't seen it has any idea of the water here, and the wind is strong enough to blow a man off his feet. Great waves come over the end of the pier, and carry everything, that's not lashed, into the sea. One day, a few winters ago, a perfect wall of water thundered down on the pier and twisted that big iron crane you see out there as if it had been made of wire. The water often comes down the chimneys of the watch-house at the end of the pier and puts out the fires; and every time the sea comes over, the whole building shakes, as if an earthquake was going on. What's worse almost than the sea is the terrible cold. Why, sir, I've seen this pier a mass of ice from end to end, and the masts and shrouds of the vessels moored alongside also covered with ice; so that a rope, which was no thicker than your finger, would look as big as a man's arm. As you know, sir, it's a hard frost that freezes salt water, and yet the lifeboat goes out in weather like that."

"It's a wonder to me," I said, "that under such circumstances the boat is manned."

"No difficulty in that, sir; there are always more men wanting to go out than there's room for. Now suppose a gun was fired at this minute from any of the lightships to tell us that assistance was needed you would see men running from every quarter, all eager for a place. I know how they would scramble across those boats, for I've seen them, and I've done it myself. Many a time have I jumped out of my warm bed in the middle of a winter night when a gun has fired, and rushed down to the harbour with my clothes under my arm; even then I've often been too late."

"What do you consider to be the best piece of service the *Bradford* has done?" was my next question.

"The rescue of the survivors of the *Indian Chief* in the beginning of 1881. The men were out for over twenty-four hours in a terrible sea and dreadful cold. I was, unfortunately, away piloting when they started, but returned in time to see them come in. Though I knew all the boatmen well, I could not recognise a single one, the cold had so altered their faces, and the salt water had made their hair as white as wool. I can never forget it. Fish, the coxswain, received a gold medal from the Institution. There was a song made about the rescue, and us Ramsgate boatmen used to sing it. When the coxswain gave up his post, about three years ago, he got a gold second service clasp, the first ever given by the Institution. In twenty-six years he was out in the lifeboat on service nearly four hundred times, and helped to save about nine hundred lives. That's the third *Bradford* we've had here. The first was presented by the town of Bradford in Yorkshire, the sum for her equipment being collected in the Exchange there in an hour. That's how she got her name, and it's been kept up ever since.

"It's no joke, I can tell you," he continued, "being out in the lifeboat. In a ship you can walk about and do something to keep yourself warm, but in the boat you've got to sit still and hold on to the thwart if you don't want to be washed overboard. Like enough you get wet to the skin before you start, and each wave that breaks over the boat seems to freeze the very blood in your veins. Then, when you reach the wreck, it is low tide, and there you've got to wait till the water rises, for in some places the sands stand as high as seven feet out of the sea when the tide is down. Then, when the lifeboat gets alongside the wreck, every man requires to have his wits about him, watching for big waves, keeping clear of the wreckage, and getting the men on board. Many a time have I gone home, after being out for six or eight hours, and taken off my waterproof, and it has stood upright on the floor as if it had been made of tin. Perfectly true, sir, it was frozen. In a day or two we forget all about the hardships we have suffered, and are as ready as ever to go out when the summons comes. We never stop to ask whether the shipwrecked men are Germans, Frenchmen, or Italians. They must be saved, and we are the men to do it. We get used to the danger in time, and think very little about it."

AN OLD WRECK.

We talked for some time longer about the treacherous nature of the Goodwin Sands, and he told me that vessels are sometimes swallowed up in a few days after they are wrecked, but occasionally they remain visible for a longer period. One large iron vessel, laden with grain, which went ashore nearly four years ago is still standing, and in calm weather the tops of her iron masts may be seen sticking out of the water.

My informant was now wanted to take charge of a party of ladies who were going out for a row, so I said "Good-bye," and came away deeply impressed with the simple heroism of the lifeboatmen, of whom this man is but a type.

CHAPTER V.
THE BOATMEN OF THE DOWNS.

There's fury in the tempest,
And there's madness in the waves;
The lightning snake coils round the foam,
The headlong thunder raves;
Yet a boat is on the waters,
Filled with Britain's daring sons,
Who pull like lions out to sea,
And count the minute guns.

'Tis Mercy calls them to the work--
A ship is in distress!
Away they speed with timely help
That many a heart shall bless:
And braver deeds than ever turned
The fate of kings and crowns
Are done for England's glory,
By her Boatmen of the Downs.

We thank the friend who gives us aid
Upon the quiet land;
We love him for his kindly word,
And prize his helping hand;
But louder praise shall dwell around
The gallant ones who go,
In face of death, to seek and save
The stranger or the foe.

A boat is on the waters--

When the very sea-birds hide:

'Tis noble blood must fill the pulse

That's calm in such a tide!

And England, rich in records

Of her princes, kings, and crowns,

May tell still prouder stories

Of her Boatmen of the Downs.

ELIZA COOK.

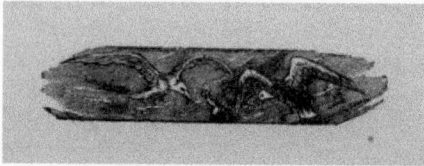

Chapter V tailpiece

CHAPTER VI.
A GOOD NIGHT'S WORK.

About a quarter past eight one wintry night, a telegram was received at Ramsgate to say that the lightships west of Margate were sending up rockets and firing guns. Owing to the rough sea and strong wind, the Margate lifeboat had been unable to leave the beach, so the coxswain decided to send news of the disaster to Ramsgate, for he knew that the lifeboat there was able, by the help of the tug, to go out in any weather.

The appeal was not made in vain, and in an astonishingly short space of time the tug and lifeboat were on their way to the Goodwins. For a long time they were unable to find out the position of the wreck, and had begun to fear that they had arrived too late, when suddenly the flare of a tar-barrel lighted up the gloom and showed them a large ship hard and fast upon the sands. The water lashed round her in tremendous surges, and every wave seemed to make her tremble from stem to stern. The boatmen at once prepared for action. The tow rope was cast off, the sail hoisted, and the lifeboat plunged quickly through the broken water.

The shipwrecked people saw her coming, and raised a joyful shout. For hours they had been expecting to meet their awful fate, as each wave rolled towards the ship, and they had prepared for death; but when they saw help so near, the love of life was once more roused within them, and they watched the boat with frantic eagerness. The sail was lowered, the anchor thrown overboard, and the cable was slacked down towards the vessel. Unfortunately, the men had miscalculated the distance, and when all the rope was run out, the boat was not within 60 feet of the wreck. Slowly and laboriously the cable had to be hauled in before another attempt could be made to get alongside. The anchor had taken such a firm hold that it required the utmost exertions of the men to raise it, but at last they succeeded. They then sailed closer to the ship, and heaved the anchor overboard again. This time they had judged the distance correctly, and after they had secured a rope from the bow and another from the stern of the ship they were ready to begin work.

The wrecked vessel was the *Fusilier*, bound from London to Australia with emigrants. She had on board more than a hundred passengers, sixty of whom were women and children. As soon as the lifeboat got near enough, the captain called out to the men in the boat, "How many can you carry?" They replied that they had a steam tug waiting not far off, and said that they would take the passengers and crew off in parties to her. As the boat rose on the crest of a wave, two of the brave fellows caught the ship's ropes and climbed on board. "Who are you?" shouted the captain as they jumped down on to the deck among the excited passengers. "Two men from the life-boat," and at these words the men and women crowded round them, all eager to seize them by the hand, some even clinging to them in the madness of their terror. For a few moments there was a scene of wild excitement on deck, and it took all the authority of the captain to restore order and quietness.

It was then arranged that the women and children should be saved first. It was indeed a task of no little difficulty, for the lifeboat was pitching and tossing in a most terrible manner. At one time she was driven right away from the ship, then back again she came threatening to dash herself to pieces against the side of the vessel, then almost at the same instant she rose on the top of a wave nearly to the level of the ship's deck.

The first woman was brought to the side, but the moment she saw the frightful swirl of waters she shrank back and declared she would rather perish than make the attempt. There was no time to waste on words. She was taken up and handed bodily to two men suspended by ropes over the vessel's side. The boat rose on a wave, and the men stood ready to catch her. At a shout from them, those who were holding the woman let go, but in her fear she clung to the arm of one of the men. In another moment she would have dropped into the sea had not a boatman caught hold of her heel and pulled her into the boat. So one after another were taken off the wreck, and soon the boat was filled. Just as the ropes were being cast off, a man rushed up to the gangway and handed a bundle to one of the sailors. Thinking that it was only a blanket which the man intended for his wife in the boat, he shouted out, "Here, catch this!" and tossed it to one of the men. Fortunately, he succeeded in catching it, and was astonished to hear a baby cry. The next instant it was snatched from his hand by the mother.

At length the anchor was weighed, the sail hoisted, and the lifeboat headed for the tug. A faint cheer was raised by the remaining passengers, who watched her anxiously as she made her way, half buried in spray, through the sea. As is often the case with those rescued from shipwreck, the emigrants thought they were safer on the wreck than in the lifeboat, and as the huge seas swept over them, they feared that they had only been saved from death in one form to meet it in another.

Soon, however, their hearts were gladdened by the sight of the tug's lights shining over the water, and in a few minutes the boat was alongside. Hastily, yet tenderly, the women were dragged on board the tug. Every moment was precious for the sake of those left behind. One woman wanted to get back to the boat to look for her child, but her voice was drowned in the roar of the storm, and she was taken below. Then, again, the bundle is tossed through the air and caught, and just as it was about to be thrown into a corner, some one shouted, "That's a baby!" It was carried down into the cabin and given to the mother. She received her child with a great outburst of joy, and then fell fainting on the floor.

The lifeboat, having discharged her load, set forth again for the wreck. All the former dangers had to be faced and all the former difficulties overcome before the work of rescue could be resumed, but the gallant fellows persevered and were successful. The boat was rapidly filled, and again made for the steamer, to which the rescued people were transferred without mishap. The third and last journey was attended with equal good fortune. All were saved--families were reunited, and friends clasped the hands of friends. Then the lifeboat went back to remain by the wreck, for the captain thought that the ship might be got off with the next high tide.

The tug with her burden of rescued people started for Ramsgate just as day was dawning. As she steamed slowly along, the look-out man noticed a portion of a wreck to which several men were clinging. At once the tug put about to bring the lifeboat to the scene. In a short time she returned with the lifeboat in tow. Having been put in a proper position for the wreck the tow rope was cast off, and the boat advanced to the battle alone. From the position of the wreck the lifeboatmen saw that the only way of rescuing the crew was by running straight into her. This was a course attended with considerable danger, but it was the only one, so the risk had to be taken. Straight in among the floating wreckage dashed the lifeboat, a rope was made fast to the fore-rigging, and the crew, sixteen in number, dropped one by one from the mast into the boat. Then the sail was hoisted, and the lifeboat made for the steamer, the deck of which was crowded with the lately-rescued emigrants, who cheered till they were hoarse, and welcomed the rescued men with outstretched arms.

The poor fellows had a touching story to tell. For hours they had clung to the mast, hearing the timbers cracking and smashing as the heavy sea beat against the wreck, and fearing that they would be swept away every minute. They had seen the steamer's lights as she passed them on her errand of mercy the night before, and had shouted to attract the notice of those on board, but the roar of the wind drowned their voices. When they saw the steamer in the morning they were filled with new hope, and made signals to attract her attention, but to their horror she turned and went back. At first they thought that they were to be abandoned to their fate, and then it dawned upon them that she had gone for the lifeboat. This was, as we know, the case. Their vessel was named the *Demerara*.

There was a scene of great enthusiasm on Ramsgate pier, when the tug, with the lifeboat in tow, entered the harbour with flags flying to tell the glad news that all were saved; and as the one hundred and twenty rescued men, women, and children were landed, cheer after cheer rent the air. It is interesting to know that the *Fusilier* was afterwards got off the sands.

CHAPTER VII.
THE "BRADFORD" TO THE RESCUE.

Of the many heartrending scenes which have taken place on our coasts, there is perhaps none more calculated to move our sympathies for the imperilled crews, and our admiration for the devotion and unconquerable courage of our noble lifeboatmen, than the wreck of the *Indian Chief*, which took place on the 5th of January 1881. The vessel stranded at three o'clock in the morning, and the crew almost immediately took to the rigging, where they remained for thirty hours exposed to the raging elements, and in momentary expectation of death. During the night one of the masts fell overboard, and sixteen unfortunate men, who had lashed themselves to it, were drowned in sight of their comrades, who were powerless to afford them any aid.

Meanwhile, word had reached Ramsgate that a large ship had stranded on the Goodwins. The tug *Vulcan*, with the lifeboat *Bradford* in tow, was accordingly sent out to render assistance. There was a strong south-easterly gale blowing, and the sea was running very high. As the boats left the harbour on their noble mission, volumes of water burst over them, and the lifeboat was frequently hidden from the gaze of the hundreds who thronged the pier to witness her departure.

The wind was piercing, and, as one of the crew afterwards declared, it was more like a flaying machine than a natural gale of wind; but it was not until they had got clear of the North Foreland that they experienced the full force of the tempest. The tug was only occasionally visible, and it seemed a perfect miracle that she did not founder. The lifeboat fared no better, for the heavy waves dashed into her as if they would have knocked her bottom out.

The short January day was now drawing rapidly to a close, and still the wreck was not in sight. What was to be done? The question was a serious one, and so the men began to talk the matter over. It was bitterly cold, and if they remained where they were their sufferings would be great; but then they would be on the spot to help their fellow-creatures as soon as another day gave them sufficient light to see where they were.

"We had better stop here and wait for daylight," said one.

"I'm for stopping," said another.

"We're here to fetch the wreck, and fetch it we will, if we wait a week," shouted a third.

Without a murmur of dissent or a moment's hesitation, the brave fellows prepared to pass the night in the open boat. But first they had to communicate with the tug. They hailed her, and when she came alongside they informed the captain of their intention. "All right," he shouted back, and then the steamer took up her position in front, keeping her paddles slowly revolving, so that she should not drift.

Throughout the night these gallant lifeboatmen lay huddled together for warmth in the bottom of the boat. In such weather it required vigorous exercise to keep the blood circulating, and before morning dawned several of the men were groaning with the cold, and pressing themselves against the thwarts to relieve the pain. But even these hardships were borne without complaint, as they thought of the sufferings of the shipwrecked crew, and jokes were not wanting to help to pass the time.

"Charlie Fish," said one of the boatmen, speaking to the coxswain, "what would some of them young gen'l'men as comes to Ramsgate in the summer, and says they'd like to go out in the lifeboat, think of this?" A general roar of laughter was the answer.

At length the cold grey light of early dawn proclaimed the advent of a new day. Keen eyes gazed anxiously towards the sands for a sight of the wreck. At first nothing was visible but tall columns of whirling spray, then after a time a mast was seen sticking up out of the water about three miles off. The scene was enough to make the stoutest heart quail, and the lifeboatmen held their breath as they looked at the water rushing in tall columns of foam more than half-way up the mast. The roar of the sea could be heard even above the whistling of the wind.

The feeling of fear, however, seems to have no place in the heart of the lifeboatman, and in a few minutes the *Bradford* was cast loose from the tug, her foresail was hoisted, and away she sped into the surf on her errand of mercy, every man holding on to the thwarts for dear life. As they approached nearer the vessel they could see a number of men dressed in yellow oilskins lashed to the foretop. The sea was fearful, and the poor fellows, who had long since abandoned all hope, were afraid that the lifeboat would be unable to rescue them. Little did they know the heroic natures of the crew of the *Bradford*. Sooner would every man have gone down to a watery grave than abandon the wreck till all were saved!

The boat came to close quarters, and the anchor was thrown out. The sailors unlashed themselves and scrambled down the rigging to the shattered deck of their once noble ship. The boatmen shouted to them to throw a line. This was done, a rope was passed from the lifeboat to the wreck, and the work of rescue began.

Where the mast had fallen overboard there was a horrible muddle of wreckage and dead bodies. "Take in that poor fellow there," shouted the coxswain, pointing to the body of the captain, which, still lashed to the mizzenmast, with head stiff and fixed eyeballs, appeared to be struggling in the water. The coxswain thought he was alive, and when one of the sailors told him that the captain had been dead four hours, the shock was almost too great to be borne. Little wonder is it that these gallant fellows were haunted by that ghastly spectacle for many a day, and it was no uncommon thing for them to start up from sleep, thinking that these wide-open, sightless eyes were gazing upon them, and the dumb lips were calling for help.

The survivors were taken off the wreck with all speed, and the boat's course was shaped for Ramsgate harbour. Outside the sands the tug was in waiting, a rope was quickly passed on board, and away they steamed. Meanwhile, news had come to Ramsgate that three lifeboats along the coast had gone out and returned without being able to reach the wreck. This naturally caused great anxiety in the town, and it was feared that some accident had befallen the *Bradford*. From early morning on Thursday, anxious wives and sisters were on the lookout on the pierhead. About two o'clock the *Vulcan* came in sight with the lifeboat astern. Almost immediately the pier was thronged with a crowd numbering about two thousand persons. At half-past two the tug steamed into the harbour, having been absent upwards of twenty-six hours.

"One by one," writes Clark Russell, "the survivors came along the pier, the most dismal procession it was ever my lot to behold, eleven live but scarcely living men, most of them clad in oilskins, and walking with bowed backs, drooping heads, and nerveless arms. There was blood on the faces of some, circled with a white encrustation of salt, and this same salt filled the hollows of their eyes and streaked their hair with lines which looked like snow. They were all saturated with brine; they were soaked with sea-water to the very marrow of their bones. Shivering, and with a stupefied rolling of the eyes, their teeth clenched, their chilled fingers pressed into the palms of their hands, they passed out of sight. I had often met men newly rescued from shipwreck, but never remember having beheld more mental anguish and physical suffering than was expressed in the countenances and movements of these eleven sailors."

SURVIVORS OF THE "INDIAN CHIEF."

They were taken to the Sailors' Home, and well cared for; the lifeboatmen were escorted home to their families amid the cheers of the spectators. Thus ended a splendid piece of service. "Nothing grander in its way was ever done before, even by Englishmen."

Five days later a most fitting and interesting ceremony took place on the lawn in front of the coastguard station at Ramsgate, when the medals and certificates of the Royal National Lifeboat Institution were awarded to those who had taken part in the rescue. The coxswain of the *Bradford* received the gold medal, each of the crew of the lifeboat and the captain of the tug received silver medals, the engineer was presented with the second service clasp, and a certificate of thanks was handed to each of the *Vulcan's* crew. The Duke of Edinburgh, himself a sailor, in distributing the honours, told the men that their heroic conduct had awakened the greatest possible interest and pride throughout England; and he declared his conviction that though they would prize the rewards greatly, they would most value the recollection of having by their pluck and determination saved so many lives.

CHAPTER VIII.
THE LAST CHANCE.

Exactly ten years after the events narrated in the previous chapter had taken place, the Ramsgate lifeboatmen were again conspicuous for their gallantry in saving life under the most trying circumstances. About one o'clock on the morning of the 6th of January 1891, the schooner *Crocodile*, bound for London with a cargo of stone, ran ashore on the Goodwins. Blinding snow squalls prevailed at the time, and the wind blew with the force of a hurricane. Immediately the vessel struck, she turned completely round and went broadside on to the sands. On realising their position, the crew burnt flares, made by tearing up their clothes and soaking the rags in oil, and attracted the attention of those on board the Gull lightship, who immediately fired signal-guns to summon the lifeboat. Scarcely, however, had the flare been burned than the sailors were compelled by the high seas to take to the rigging. Great waves swept the decks, carrying everything before them; even the ship's boats were wrenched from the davits and whirled away as if they had been toys.

In answer to the guns the Ramsgate tug and lifeboat were manned and steered in the direction of the flare. Huge seas broke over the lifeboat and froze as they fell on the almost motionless figures of the boatmen. The snow came down in pitiless showers, enveloping them in its white mantle. In a short time the tug had towed the *Bradford* to windward of the vessel. Then the rope was thrown off, the sail was hoisted, and the boat made for the wreck. She had not gone far before a terrific snow squall overtook her. Fearing that they would be driven past the vessel without seeing her, the coxswain ordered the anchor to be thrown out. This was done, and the boat lay-to till the sudden fury of the gale had spent itself. Then the anchor was hoisted in and all sail made for the wreck.

Again the anchor was let go, just to windward of her, and the lifeboat was veered cautiously down. As they drew nearer, the men could see the crouching figures of the sailors lashed to the rigging. They seemed more dead than alive, and gazed upon the men who were risking their own lives, to save them with the fixed stare of indifference or death. The lifeboat ran in under the stern and was brought up alongside. The grapnel was got out, and one of the men stood up, ready to throw it into the rigging on the first favourable opportunity. Suddenly a mighty billow swooped down upon them. The anchor cable--5 inches thick--was snapped like a thread, and the boat was borne on the crest of the wave far out of reach of the wreck.

A LIFEBOAT GOING OUT.

A LIFEBOAT GOING OUT.

As quickly as possible the sail was again set, and the trusty *Bradford* made for the tug, which was burning blue lights to show where she was. After many attempts a rope was secured on board, and the *Aid* steamed to windward the second time with the lifeboat in tow. Once more she was in a favourable position for the wreck, the rope was cast off, and the sail hoisted. The second and last anchor was let go, and the cable was slowly slackened. If they failed this time the men must perish. It was a terribly anxious moment, but fortune favoured them, and the lifeboat was successfully brought into her former position alongside.

The hull of the *Crocodile* was now entirely under water, and her deck was washed by every wave. High up in the rigging, on the side opposite to that on which the lifeboat lay, the crew were huddled. The only way for them to reach the lifeboat was by climbing to the masthead and coming down on the other side. This is a feat which requires no little steadiness of hand and eye, and when we remember that these poor sailors had been exposed for nearly five hours on this January night to the full fury of a wintry storm, we shall be better able to appreciate the terrors through which they passed before they found themselves safe in the lifeboat.

In obedience to the coxswain's order, they unlashed themselves and began to crawl aloft. Every sea shook the vessel, and, as she settled again on the sands, the masts bent almost double. Their progress was slow, but before long they were in a position to be rescued. This was done with great difficulty, for the heavy seas caused the lifeboat to strike against the vessel several times with considerable violence, but her cork fender protected her from injury. At length the whole crew of six men were hauled safely on board. The captain alone remained to be rescued.

High up at the masthead he could be seen preparing to cross from the opposite side. Benumbed by the cold and bewildered by the swaying of the masts, he paused for a moment. The lifeboatmen shouted words of encouragement to him, and he prepared to come on, but he missed his hold and fell into the seething waves eddying round the wreck. As he fell his lifebelt caught on something, and was torn off, and before the boatmen could lay hold of him he was swept out of their sight for ever.

The lifeboat was quickly got clear of the wreck, and proceeded under sail to the tug, which was in waiting some distance off. Ramsgate was reached about eight o'clock in the morning, where the rescued men were supplied with dry clothing and food, of which they stood greatly in need.

There is a circumstance of peculiar interest connected with the wreck of the *Crocodile*. Two days before she struck on the sands, her sister ship, the *Kate*, also laden with stone, was stranded on the Goodwins. On that occasion the lifeboat *Mary Somerville* of Deal went out to assist. The lifeboatmen were employed to throw the cargo overboard and try to get the vessel afloat. This was successfully accomplished, and on the morning of the day on which the *Crocodile* was wrecked, her sister ship was towed into Ramsgate harbour with her crew of nine men on board.

CHAPTER IX.
HARDLY SAVED.

The first duty of the crew of the lifeboat is to save life, but it frequently happens that a stranded vessel is not so seriously damaged as to hinder her being got afloat again. Under these circumstances the men are at liberty to assist in saving the vessel if the captain is willing to employ them. This is a very dangerous business, and often after long hours of peril and labour the ship is dashed to pieces by the waves, and the men are with difficulty rescued. A splendid example of the risk attending this salvage service occurred several years ago on the Goodwin Sands.

In response to signals of distress the tug and lifeboat put out from Ramsgate pier, and found a Portuguese ship on the sands. Her masts and rigging were still standing, and there was every chance of her being saved. The vessel had gone head on to the Goodwins, and the boatmen got an anchor out from the stern as quickly as possible, with the intention of working her off into deep water by the help of the tug; but this attempt had soon to be abandoned. Shortly after midnight the gale increased, and heavy seas began to roll over the sands. The ship, which had all along lain comparatively still, was now dashed about by the waves with terrific violence. The lifeboat remained alongside, and her crew, knowing well that a storm on the Goodwins is not to be trifled with, urged the sailors to come on board. The captain, however, refused to leave his ship, so there was nothing for it but to wait until an extra heavy sea should convince the captain that it was no longer possible to save the vessel.

This happened sooner than could have been expected, for almost the very next instant a wave struck her and smashed several of her timbers. The sailors now begged to be taken on board, and they were told to "Come on, and hurry up." But first of all they had to get their belongings. Though every moment was of consequence, the coxswain had not the heart to forbid them bringing any articles on board, and eight chests were lowered into the lifeboat. Then one by one the crew abandoned the vessel.

All danger was not yet over. The seas dashed over the ship into the lifeboat, blinding and drenching the men, and rendering still more difficult their task of keeping the boat from being crushed under the side of the vessel. Haul at the cable as they would, they were unable to get her out of the basin which the brig had made for herself in the sand. To add to the horror of their position, the wreck threatened to fall over on the top of them every moment.

There was only one way of escape--to wait until the tide rose sufficiently to float them off, but the chances were that when the tide rose it would be too late to save them. They would then have ceased to struggle or to suffer, and the battered remains of their trusty boat would tell those at home what had become of them. Crouching down as low as possible to avoid being struck by the swaying yards and fluttering canvas, the men waited for deliverance--or would it be death?

At length the tide reached her, and the boatmen redoubled their efforts to haul their little vessel away from the ship. Slowly, very slowly, she drew away from that terrible black hull and those swaying yards. But now a new and unforeseen difficulty presented itself. In the face of the wind and tide it was impossible for them to get away from the sands, so in spite of their exhaustion and the black darkness of the night, they determined to beat right across the sands. They hauled hard on the cable again, but the anchor began to drag, and they were drifting back again to the wreck.

"Up foresail!" shouted the coxswain, at the same time giving orders to cut away the anchor. The boat bounded forward for a few yards and then struck on the sands again fearfully near to the wreck. Wave after wave dashed into the boat and nearly washed the wearied men overboard, but they held on like bulldogs. Three times she was driven back to the wreck, and again and again she grounded on the sands.

One of the crew, an old man upwards of fifty years of age, thus described his feelings.

"Perhaps my friends were right when they said I hadn't ought to have gone out, but, you see, when there is life to be saved, it makes a man feel young again; and I've always felt I had a call to save life when I could, and I wasn't going to hang back then. I stood it better than some of them, after all; but when we got to beating and grubbing over the sands, swinging round and round, and grounding every few yards with a jerk, that almost tore our arms out from the sockets; no sooner washed off one ridge, and beginning to hope that the boat was clear, than she thumped upon another harder than ever, and all the time the wash of the surf nearly carrying us out of the boat--it was truly almost too much for any man to stand. I cannot describe it, nor can anyone else; but when you say that you've beat and thumped over these sands, almost yard by yard, in a fearful storm on a winter's night, and live to tell the tale, why it seems to me about the next thing to saying that you've been dead and brought to life again."

At length deep water was reached, and their dangers were over. Quickly more sail was hoisted, and the boat headed for the welcome shelter of Ramsgate pier. All were in good spirits now, even the Portuguese sailors who had lost nearly everything they possessed. On the way home the lifeboatmen noticed that they seemed to be discussing something among themselves. Presently one of them presented the coxswain with all the money they could scrape together, amounting to about £17, to be divided among the crew. "We don't want your money," shouted the hardy fellows, and with many shakings of the head they returned the generous gift. The harbour was soon afterwards reached, where they were landed overjoyed at their miraculous escape, and by every means in their power endeavouring to show the gratitude they felt but could not speak.

CHAPTER X.
A WRESTLE WITH DEATH.

One bleak December night, a few years ago, word was brought to Ramsgate that a large vessel had gone ashore on the Goodwin Sands. Immediately on receiving the message, the harbour-master ordered the steam tug *Aid* to tow the lifeboat to the scene of the disaster. The alarm bell was rung, the crew scrambled into their places, a stout hawser was passed on board the tug, and away they went into the pitchy darkness.

The storm was at its height, and "the billows frothed like yeast" under the lash of the furious wind. Hardly had the lifeboat left the shelter of the breakwater than a huge wave burst over her, drenching the men to the skin, in spite of their waterproofs and cork jackets, and almost sweeping some of them overboard. At one moment they were tossed upwards, as it seemed to the sky; at another they dropped down into a valley of water with huge green walls on either side. Again and again the spray dashed over them in blinding showers, but no one thought of turning back.

Bravely the stout little tug battled with the waves, and slowly but surely made headway against the storm, dragging the lifeboat after her. As they neared the probable position of the wreck, the men eagerly strained their eyes to gain a sight of the object of their search, but nothing met their gaze save the white waters foaming on the fatal sands. Suddenly, through the flying spray, loomed the hull of a large ship, with the breakers dashing over the bows. Not a single figure was visible in the rigging, and on that desolate, wave-swept deck no mortal man could keep his footing for five seconds.

"All must have perished!" Such was the painful conclusion arrived at by the lifeboatmen as they approached the stranded vessel, but it would never do for them to return and say that they *thought* all the crew had been swept away; they must go and find out for certain. The tow rope was accordingly thrown off, the sail was hoisted, and the lifeboat darted among the breakers. Suddenly one of the lifeboatmen uttered a cry, and on looking in the direction of his outstretched arm, his companions saw four figures crouching under the lee of one of the deck-houses. The anchor was immediately let go, and the lifeboat was brought up under the stern of the wreck.

To the astonishment of the boatmen the sailors had as yet hardly noticed their presence. They seemed to be deeply absorbed in making something, but what it was could not be seen. Presently one of the men rose up, and coming to the stern of the vessel threw a lifebuoy attached to a long line into the sea. It was afterwards learnt that, from the time their vessel struck, these poor fellows had busied themselves in preparing this buoy to throw to their rescuers when they should arrive.

Borne by the wind and tide the lifebuoy reached the boat, and was at once seized and hauled on board. An endeavour was then made to pull the lifeboat nearer the wreck, but the strength of the men was of no avail against that of the tempest. Great seas came thundering over the wreck and nearly swamped the boat. Several men were shaken from their places, but fortunately none of them were washed overboard. They redoubled their efforts after each repulse, but with no better fortune.

Seeing that the lifeboat could not come to him, the captain of the doomed vessel determined to go to her. Choosing a favourable moment, he abandoned the shelter of the deck-house, threw off his coat, seized hold of the line, and jumped into the sea. The waves tossed him hither and thither as they would a cork, but he held on like grim death. At one moment he hung suspended in mid air; at another he was engulfed by the raging waters. The lifeboatmen, powerless to render any assistance, watched the unequal contest with bated breath. Bravely the captain struggled on, and gradually reduced the distance between himself and the hands stretched out ready to save him. Suddenly a tremendous wave broke over the wreck, and when it passed the men saw that he had been swept from the rope.

With all the might of his strong arm the coxwain hurled a lifebuoy towards the drowning man. Fortunately it reached him, and with feelings of inexpressible relief the men saw him slip his shoulders through the buoy as he rose on the crest of a breaker. "All right," he shouted, as he waved his hand and vanished in the darkness.

Suddenly a terrific crash reminded the lifeboatmen that there were still two men and a boy on the wreck. Turning round they saw that the mainmast had given way and gone crashing overboard. Startled by the suddenness of the shock the survivors supposed that the end had come, and with a blood-curdling scream of despair they rushed to the side of the vessel imploring aid. The chief mate sprang into the water and endeavoured to swim to the lifeboat. The men again laid hold of the rope and tugged with might and main to get nearer the wreck, but the storm mocked their efforts. Then they tried to throw him a line, but it fell short. Again and again they tried, but in vain. The mate battled bravely for life, and as he was a powerful man, all thought that he would succeed, but he was weakened by exposure and want of food, and his strength was rapidly failing. The lifeboatmen exerted themselves to the utmost to reach him, pulling at the rope till every vein in their bodies stood out like whipcord. Not an inch could they move the boat. The man's agonising cries for help nearly drove them mad, but they could do no more. His fate was only a matter of time, and in a few moments he sank into his watery grave, with one long shriek for help.

There were still a man and a boy on the wreck. With heavy hearts, and a dimness about the eyes that was not caused by the flying spray, the lifeboatmen once more vainly attempted to get nearer the wreck. Following the captain's example, the man seized the rope and jumped into the water. Fortune favoured him, and though he was tossed about in a frightful manner he succeeded in pulling himself right under the bows of the lifeboat. Then his strength failed, and he would have been instantly swept away and drowned, had not one of the lifeboatmen flung himself half-way over the bow of the boat and caught the perishing sailor by the collar. Stretched on the sloping foredeck of the boat he could not get sufficient purchase to drag the man on board, and indeed he felt himself slowly slipping into the sea.

"Hold me! hold me!" he cried, and several of his companions at once seized him by the legs. The weight of the man had drawn him over till his face almost touched the sea, and each successive wave threatened to suffocate him. To add to the horror of the situation, a large quantity of wreckage was seen drifting right down upon the bow of the boat towards the spot where the men were struggling. If it touched them it meant death. For a moment it seemed endued with life, and paused as if to consider its course, then just at the last minute it spun round and was borne harmlessly past.

The crew now made a desperate attempt to haul the two men on board. Finding that the height of the bow prevented their success, they dragged them along the side of the boat to the waist, and pulled them in wet and exhausted.

The boy alone remained on the wreck, which was now fast breaking up. How to help him was a question not easily answered, for with all their pulling they could not approach nearer the vessel. Suddenly the difficulty was solved for them in a most unexpected manner. A tremendous sea struck the vessel and swept along the deck. When the spray cleared away the boy was nowhere to be seen. Anxiously every eye watched the water, and presently a black object was seen drifting towards the boat. "There's the boy!" shouted the men in chorus. Slowly, very slowly, as it seemed to them, he drifted nearer and nearer. At length he came within reach of a boat-hook, and was lifted gently on board--unconscious, but still alive. After the usual restoratives had been applied, he revived.

SAVING THE CAPTAIN.

Nothing more could be done at the wreck now, so the sail was hoisted and the boat's head turned towards the harbour. But their work of saving life was not yet done. As they sped along before the blast a dark object was seen tossing up and down upon the waves. They steered the boat towards it, and to their astonishment found the captain with the lifebuoy round him, still battling for life. He was hauled on board in an utterly exhausted condition. Before reaching the shore he revived, and told the men that his vessel was the *Providentia*, a Finland ship, and that he himself was a Russian Finn. The men were landed at Ramsgate in safety. A few days later, news came from Boulogne that the remainder of the crew, who had left the wreck in a boat, had been blown across the Channel and landed on the French coast.

CHAPTER XI.
A DOUBLE RESCUE.

Clang! clash! roar! rings out the bell at the lifeboat-house, its iron voice heard even above the thunder of the surf and the whistling wind, warning the sleeping inhabitants of Deal that a vessel has gone ashore on the Goodwins. A ray of light gleams across the dark street as a door opens and a tall figure rushes out--it is that of a lifeboatman. Presently he is joined by others, and all hurry on as fast as possible, in the face of the furious wind, to reach the boathouse. Each man buckles on his lifebelt, and takes his place in the lifeboat. Those who have failed to get a place help to run it down to the white line of surf, over the well-greased boards laid down on the shingle. The coxswain stands up in the stern with the rudder lines in his hands, watching for a favourable moment to launch. The time has come, the order is given, and away dashes the lifeboat on her glorious errand.

Onward she plunged under close-reefed sail in the direction of the flares, which the shipwrecked men were burning to tell the rescuers of their whereabouts. Suddenly the light went out and was seen no more. A shriek echoed over the waves, but none could say whether it was that of "some strong swimmer in his agony," or only the voice of the wind. The lifeboatmen looked around them on every side, but they could see nothing; they listened, and heard nothing; they shouted, but no answer came back. "A minute more and we would have had them," says the coxswain. "Hard lines for all to perish when help was so near."

Suddenly, through the darkness, the light of another flare was seen. The boat was at once brought round and headed for the newly-discovered wreck. It was now midnight, and the sea was like a boiling cauldron, but the fine seamanship of the crew was a match for the storm. Many an anxious glance was cast in the direction of the flare, and a fervent hope was in every heart that this time they would not be too late.

"Hullo! what's that?" exclaimed the lifeboatmen together, as a dark object rose in the sea between them and the flare. Another wreck! And sure enough there lay the dismasted hull of a large ship tossing helplessly about from side to side, with the waves dashing over her in spiteful fury. "Let us save the poor fellows," said the lifeboatmen. The anchor was let go, and the boat veered down to the stern of the wreck. Then began the tug of war. "What pen can describe the turmoil, the danger, and the appalling grandeur of the scene, how black as Erebus, and again illumined by a blaze of lightning? And what pen can do justice to the stubborn courage that persevered in the work of rescue, in spite of the difficulties which at each step sprang up?"

The shipwrecked crew were Frenchmen, and all efforts to make them understand what was wanted of them were in vain. As they crawled along the deck to the stern of the vessel they presented a most pitiable sight, and when the lifeboatmen shouted to them to "come on and take our line," they paid no attention. Suffering and exposure seemed to have deprived them of their mental faculties. Time after time a line was thrown to them, but they allowed it to slip back into the sea, without attempting to lay hold of it. Then the boatmen saw that if these men were to be rescued, it would be by their own unaided exertions.

How the rescue was to be effected was quite another matter, but there is never a difficulty which cannot be overcome by persistence and courage. So thought the lifeboatmen, as their boat was tossed about in that swirl of angry waters. At one minute she was swept right away from the wreck, while at another she was driven onwards and lifted upwards by a wave, till her keel touched the deck of the half-sunk vessel, from which she withdrew with a horrible grating sound. How she came through the terrible ordeal of being thrown up on the wreck time after time was a marvel, and is a splendid proof of the strength of the lifeboat.

All this time the Frenchmen stood at the stern of the ship eager for deliverance, but unable through fear to take any measures to accomplish it. Time was precious. Delay might mean death to those on the other vessel, so one of the lifeboatmen, named Roberts, hit upon a desperate plan for getting the crew off. Cautiously he crawled forward and took up his position on the fore air-box of the lifeboat. Now this air-box has a rounded roof, and therefore the task that Roberts set himself was one of no little difficulty, and to carry it out successfully required no ordinary amount of nerve.

Held by the strong arms of his companions he waited till the boat was carried towards the vessel, then he shouted to the sailors' to "come on!" At last they understood, and one after another they sprang into the arms stretched out to save them. Five men were taken off in this way, and as that seemed to be all that were on board, the anchor was hoisted in, the sail was set, and the lifeboat made for the other wreck, which was still showing signals of distress. So convulsive had been the grip of these five men, that Roberts' arm and chest were black and blue, and those marks of their desperation and his bravery the gallant boatman carried about with him for many a day.

It was now four o'clock in the morning, the men were ready to drop from fatigue, and the boat was seen to be much lower in the water than usual, even though she had five extra men on board. But "courage mounteth with occasion," and they forgot their weariness and the danger in the prospect of saving fellow-creatures from the watery grave which yawned around them.

At length the wreck was reached, and proved to be that of a Swedish vessel. The anchor was let go, and the lifeboat veered down as close as was prudent. Fortunately there was an English pilot on board, who knew exactly what the lifeboatmen wanted. Under his directions lines were passed from the wreck, and the crew were speedily taken on board the boat. The captain had his wife with him, and it was with the utmost difficulty that she could be persuaded to enter into the lifeboat, which, owing to the battering it had received at the French wreck, was almost full of water. The entreaties of her husband and the boatmen at last prevailed, and she was taken on board. Then the captain followed.

No time was now lost in weighing the anchor and setting sail for home. Slowly the lifeboat made headway against the storm, as if she was wearied and fain would rest. Just as the wintry sun glinted across the sea, the keel grated on the beach at Deal. Out sprang the lifeboatmen and dragged her into shallow water, with her burden of five Frenchmen and twelve Swedes, who were heartily welcomed, and taken where warmth and comfort awaited them.

On examination it was found that there was a hole in the bow of the boat into which a man could creep, and both her fore and aft air-boxes were full of water. Had it not been that she had still a good supply of buoyancy from the air-chambers ranged along the sides, our story would have had a far from pleasant ending. Though the boatmen had succeeded in saving seventeen lives, they were sadly disappointed that the ship to whose assistance they were summoned, had gone down so suddenly. It was not, however, any fault of theirs, for no time had been wasted in going to the rescue.

CHAPTER XII.
DEAL MEN TO THE RESCUE.

About ten o'clock on the night of the 11th of February 1894, signals of distress were observed from the Gull lighthouse by the look-out on Ramsgate pier. In response the lifeboat *Bradford* was manned; but on this occasion she was found to be hard and fast on a sandbank in the harbour. The boatmen and those on the pier exerted themselves to the utmost to get her off, but it was not till eleven o'clock that she was able to proceed to sea, in tow of the tug *Aid.* She was then too late to render any assistance.

In the meantime the signals from the lightship had been seen at Deal, a few miles farther south. The boathouse bell was rung, there was a fierce rush of men for the cork lifebelts hanging round the walls, and ten minutes later the lifeboat *Mary Somerville* was manned and launched. Away she flew before the heavy south-westerly gale, with Roberts, the coxswain, at the helm, and was soon lost to sight in the darkness. The vessel in peril was the *Franz von Matheis*, a German schooner, bound from Sunderland to Portsmouth with a cargo of coal. She kept burning flares till the lifeboat got alongside. Then the men found that she was dragging her anchors and heading rapidly towards the Goodwins.

With great difficulty the *Mary Somerville* shot under the lea of the vessel, and several of her crew jumped on board the ship, which had become unmanageable, owing to the stress of weather. The presence of the lifeboatmen put fresh strength into the exhausted muscles of the crew, and all worked together with a will in the hope of saving the vessel; but it was found impossible for lifeboatmen or crew to move about on the schooner without sustaining injury. One of the men was thrown to the deck by a terrific lurch, and had his head cut open, and every moment increased the peril. The captain therefore decided to abandon the vessel, and he, with the crew of six, were taken into the lifeboat.

Even then the danger was not over. The terrific sea and wind caused the vessel to roll tremendously. One of her yards caught the mizzenmast of the boat, and broke the fastening which kept it in its place. Down fell the mast, striking the second coxswain on the head, and knocking him insensible to the bottom of the boat. For close upon an hour the gallant fellows battled with the tempest, straining every nerve to get clear. It indeed seemed as if they and the men they had with them would never again return to shore. Each wave drove the boat against the side of the vessel with a horrible, grinding crash. The steering-yoke was broken, and the boat-hook was snapped in two, "as you would the stem of a clay-pipe between your fingers." In trying to ward off the vessel four oars were smashed, and then the men found that their boat was being held down under the ship's broadside. While in this position, the tiller, which had taken the place of the steering-yoke, was sprung, a dozen or more of her stout mahogany planks were started, and her cork fender was torn to pieces.

At last they cleared the vessel, and as it was impossible, owing to the fury of the gale, to return to Deal, they made all sail for Ramsgate harbour. Here they landed the rescued men at a quarter-past one in the morning. During the day the *Mary Somerville* was taken back to Deal. No more vivid picture of the perils through which the lifeboatmen passed could be desired than that of the bruised and battered lifeboat, as she lay high and dry in the boathouse that afternoon. The *Franz von Matheis* seems afterwards to have got a firm hold, for she remained riding at anchor very close to the sands. At daybreak next morning a tug was seen endeavouring to take the abandoned ship in tow, and about four o'clock in the afternoon she was brought into Ramsgate harbour.

CHAPTER XIII.
THE WRECK OF THE "BENVENUE."

The ship *Benvenue* of Glasgow was being towed through the Straits of Dover on Nov. 11th, 1891, when a terrible gale sprang up. Arriving off Sandgate, the vessel became quite unmanageable, and it was decided to lie-to and wait until the fury of the storm had passed. Two anchors were accordingly let go, but these, even with the assistance of the tug, were not powerful enough to hold her. Nearer and nearer to the shore she drifted. Then with a tremendous lurch she struck and began to settle down. Fifteen minutes later she foundered.

The crew were ordered to go aloft as quickly as they could, for in the rigging lay their only chance of safety. The men promptly obeyed, and secured themselves with lashings; some of them got into the topsail yards, and fastened themselves in the sails. A rocket was sent up before the ship went down, to tell those on shore that help was needed, and soon an answering streak of flame shot across the sky. Though they were in such a perilous position, the men were not at all excited, but watched with eager eyes the movements of the people on the beach.

The day wore on, and still no help arrived. Several of the crew unlashed themselves and came down from the rigging, with the intention of swimming ashore. Such an attempt was useless in the terrific sea that was running, but they all had lifebelts on, and were determined to overcome the danger. Bravely they battled for life amid the seething waters, but it was in vain. One poor fellow was seen swimming about with blood trickling down his face. He must have been dashed against the ship's rail. A mighty wave came thundering down, for a moment he was visible upon its foamy crest, and then he disappeared for ever. Another man succeeded in getting half-way to the shore, when he was seen to throw up his arms, and the waters closed over him. All who made the attempt shared a similar fate.

A PERILOUS REFUGE.

The sea was now close up to the mizzentop where the survivors were standing, and every moment they expected that the mast would go by the board. With the setting of the sun the hope of being rescued, which had buoyed them up throughout the weary hours of that long day, died out, and their spirits sank to the depths of despair. They were almost perished with cold and faint with hunger, and as no help came they gave themselves up for lost.

What were the lifeboatmen doing all this time? Surely they were not going to let fellow-creatures perish without an effort to save them? No! Early that morning the lifeboat had put off from Sandgate to the assistance of the *Benvenue*, but such terrific seas were encountered that she was driven back to the shore. As it was considered impossible to launch again at Sandgate, the boat was put on the carriage and conveyed to Hythe.

At half-past nine she was launched, manned by a crew of twenty men. The sea was, however, heavier than that experienced at Sandgate, and before the boat could get clear of the surf, she was struck by a heavy wave and capsized. The whole of her crew with the exception of three men, were thrown into the water. Nineteen of them managed to reach the land, but the other poor fellow lost his life in the raging breakers. The boat was then brought ashore and replaced on the carriage. Though repulsed, the lifeboatmen were not beaten, and they remained by their boat all day, ready to launch on the first favourable opportunity. It was not, however, until half-past nine at night, exactly twelve hours since the second attempt had been made, that their patience was rewarded. Then, as the sea had considerably moderated, it was decided to make another attempt to rescue the shipwrecked crew.

With the utmost difficulty the boat was got off, and for a time failure seemed certain. The gallant lifeboatmen persevered, and, bending to the oars with all the strength of their muscular arms, won the victory. The ship was reached, and the twenty-seven survivors, out of the crew of thirty-two men, were taken into the lifeboat. They had watched with eager eyes the almost superhuman efforts that were being made on their behalf, and when they found themselves safe on board, the pent-up feelings of many found vent in tears.

The scene on the landing of the lifeboat at Folkestone baffles description. Thousands of people had assembled at the harbour, and as soon as the boat appeared, cheer after cheer was raised, and rescuers and rescued were quickly brought ashore. The former received the hearty congratulations of everyone. The latter appeared too exhausted to bear the excitement of the moment, so they were at once conducted to a place where they received the care they needed after their exposure to the wind and waves.

Next morning the crew wrote a letter of thanks to all who had taken part in their rescue, in the following terms, touching in their simplicity,--

"We desire to tender our heartfelt gratitude for the way in which we have been rescued and cared for by the crew of the lifeboat, and the others who assisted in our rescue."

At noon a special service of thanksgiving was held in the parish church, Folkestone, and as the men bad lost all their belongings, a collection was made on their behalf.

CHAPTER XIV.
THE STRANDING OF THE "EIDER."

On the night of Sunday the 31st of January 1892, the North-German Lloyd liner *Eider*, bound from New York to Southampton, stranded on a reef of rocks off the Isle of Wight. A dense fog prevailed at the time, and a very rough sea was running. Signal rockets were immediately sent up, and about eleven o'clock the Atherfield lifeboat proceeded to her assistance. There was no immediate danger to the passengers and crew, so the captain decided to telegraph for steam tugs. The telegrams were accordingly handed into the lifeboat, and she returned to the shore to send them off.

At daylight next morning signals were made by the *Eider*, and the lifeboat again went out, and found that the captain wished to land some of the mails, and they were therefore brought ashore. Meanwhile news of the stranding of the steamer had been sent to the lifeboat stations at Brighstone Grange and Brooke, and these lifeboats at once put off and made for the scene of the disaster with all speed. The captain of the *Eider* then decided that it would be best to land the passengers, and during the day the lifeboats made altogether eighteen trips to the ship, and safely landed two hundred and thirty-three passengers, besides specie and mails. Darkness, however, came on and put an end to the work.

The next day eleven journeys were performed by the lifeboats, and one hundred and forty-six people were brought to land without accident. During Wednesday and Thursday the boats were engaged in bringing ashore bars of silver, specie, the ship's plate, and passengers' luggage. Forty-one journeys in all were made by the gallant lifeboatmen, who worked hard and nobly, and rescued three hundred and seventy-nine persons. The captain and several of the crew remained on board, and the vessel was eventually towed off the rocks and safely berthed in Southampton docks.

In recognition of the devotion to duty and self-sacrifice shown by the lifeboatmen in the work of rescue, the Emperor of Germany presented each of the coxswains of the three lifeboats with a gold watch bearing His Majesty's portrait and initials. The institution also awarded the second-service clasp to the coxswain of the Atherfield lifeboat, the silver medal to the coxswain of the Brighstone Grange lifeboat, and the third-service clasp to the coxswain of the Brooke lifeboat.

We reproduce the following poem on the stranding of the *Eider*, by special permission, from *The Star*.--

The *Eider* rode on the open sea
With her safety in God's own hand
For a thousand miles--ay, two, and three,
With never a sight of land.

A shell of steel on the world of waves
That severs the hemispheres,
That covers the depths of a thousand graves
And the wrecks of a hundred years.

She bore, unhurt, through the storm-god's din,
Through shower, and shade, and sheen,
With the death without and her lives within,
And her inch of steel between.

From the port behind, to the port beyond,
With never a help or guide,
Save the needle's point and the chart he conned,
The master has fought the tide.

On the bridge, in the Sunday twilight dim,
He has taken his watchful stand;
And he hears the sound of a German hymn,
And the boom of a brazen band.

He looks for the lights of the royal isle,
Ahead, to left, and to right;
Below there is music and mirthful smile,
For land must be soon in sight.

In sight? Not yet! for a fog creeps round
And the night is doubly dark.
"Slow speed! Hush! is it the fog-bell's sound,
Or the shriek of the siren? Hark!"

The fog-bell clangs from its seaward tower,
And the siren shrills in fear;
But the vapours thicken from hour to hour,
And the master cannot hear!

On the seaward headland, the beacon's blaze
Like a midday sun would seem,
But its warning rays are lost in the haze,
And the master sees no gleam!

"How goes the line? There is time to save!"
"It is ten fathom deep by the log."
"We have not tarried for wind or wave,
We cannot wait for the fog."

On, on! through the dark of a double night;
On, on--to the lurking rock!
No sound, no gleam of a saving light
Till the *Eider* leaps to the shock.

All night she bides where the sea death hides,
And her passengers crowd her deck;
While the leaping tides laugh over her sides
And sink from the stranded wreck.

The *Eider* has gold, she has human lives;

But these can assist no more.

Pray, pray, ye German children and wives,

For help from the English shore!

A signal is sent, and a signal is seen,

And a lifeboat--ay, two, and three,

From the shore to the vessel their crews row between,

And fight with the stormy sea.

They fight day and night, as true Englishmen can,

'Mid the roar of the storm-lash'd waves;

And the *Eider's* four hundred are saved to a man

From the terror of sea-bed graves.

The *Eider* bides, all broken and bent;

With the tide she shivers and starts,

And stands--for a time--as a monument

Of the courage of English hearts.

But longer lasting, the memoried grace

Of a noble deed and grand

Will knit the hearts of the English race

To the hearts of the Fatherland!

CHAPTER XV.
THE WRECK OF THE "NORTHERN BELLE."

During a dreadful storm which swept over the British Isles several years ago, the American ship *Northern Belle*, from New York to London, came to anchor off Kingsgate, near Broadstairs, about a mile from the shore. The sea made great breaches over her, and, in order to lighten the vessel and help her to ride out the storm, the crew cut away two of the masts. With the flood-tide, however, the gale increased, and it was feared that the vessel would drag her anchors and come ashore. A swift-footed messenger was accordingly despatched to summon the Broadstairs lifeboat.

Without delay the crew were mustered, and the boat, on her carriage, was dragged overland to Kingsgate, a distance of two miles. It was nine o'clock when the *Mary White* arrived, and by that time the cliffs were lined with crowds of people. Shortly afterwards two luggers were seen bearing down upon the unfortunate vessel. One of these crafts, when trying to take out one of the ship's anchors, was overwhelmed by a heavy sea, and sank. Not one of her crew of nine men were ever seen again. The other was more successful, and five of her crew managed to get on board the *Northern Belle*. Every moment the multitude of spectators expected to see the vessel run ashore and be dashed to pieces on the rocks at the foot of the cliff; but as the day wore on and the anchors still held, it was thought that she would yet be safe. Heedless of the heavy snow and bitter cold, the people watched her till darkness came on and shut out the vessel from their gaze.

THEY BENT THEIR BACKS TO THE OARS.

About midnight, the long-expected catastrophe took place, the cable broke and the vessel was driven on the rocks. In the storm and darkness it would have been worse than useless to launch the lifeboat, so the men were reluctantly compelled to put off the rescue till a new day should give them sufficient light to see what they were doing. Next morning, about seven o'clock, the remains of the ill-fated ship could be seen, and lashed to the only remaining mast were the figures of twenty-three perishing sailors. What they must have suffered in the cold and darkness of that terrible night may be imagined, but it cannot be described.

The lifeboat was dragged down to the water's edge, and the crew got into their places. The coxswain stood up in the stern, grasping the yoke lines, and watching for a favourable moment to put off. The faces of the men were grave, for they knew the terrific struggle that was before them, and, with such a high sea running, who knew if they would come back again? The coxswain gave the word, and the boat was pushed off into the raging surf. The boatmen bent their backs and made headway in spite of the storm. Over and over again they were lost to sight, and those on shore were filled with fear for their safety, but the good boat breasted each wave gallantly, and quickly drew near to the wreck.

Great difficulty was experienced in getting alongside, and in the struggle the bow of the lifeboat was badly damaged, but at last the boat was made fast. The poor sailors were so benumbed by their long exposure to cold that they were almost helpless, and this made the task of the boatmen still more difficult. At length, after tremendous exertions, they succeeded in taking off seven of the crew. On account of the broken condition of the boat and the high sea, it was not judged prudent to take more, so she was cut adrift from the wreck and returned to the shore with her precious burden.

Fearing that an accident might happen to the *Mary White* and disable her for further service, a second lifeboat had been brought over from Broadstairs. She was now launched, and made for the wreck, from which she shortly afterwards returned with fourteen men. Only two sailors now remained on board, the aged captain and the pilot. The former stubbornly refused to leave his ship, declaring that he would rather be drowned; and the latter said that he was not going to leave the old man to perish by himself.

The coxswain allowed two hours to pass, expecting that the captain would change his mind and signal for them to come and take him off; but when he showed no signs of yielding, he called the men together and launched the lifeboat. After a stiff pull they reached the wreck, and tried to persuade the captain to save himself, but he remained obstinate. Then the men declared that they would remain by the wreck as long as she held together, even if they waited a week. The coxswain pointed out to the captain that he was not only throwing his own life away for no good reason, but that he was also endangering the lives of those in the boat, and he told him that it was his duty to save himself. At length he was persuaded of the folly of his action, and came down from the rigging. The pilot, whose chivalrous feelings alone had kept him in this perilous position, also gladly entered the saving boat.

Great were the rejoicings on the beach when it became known that the whole crew had now been rescued. The shipwrecked men were taken to a house near at hand, but they were so exhausted that they were unable to eat.

Shortly afterwards three horses were harnessed to the transporting carriage of the *Mary White*, and she was taken back to Broadstairs. As she approached the town, the people came out to meet her, and with cheers loud and long welcomed the heroes home.

An eye-witness of the rescue says: "The lifeboatmen were not labouring under any species of excitement when they engaged in the perilous duty, which they performed so nobly and so well. Under the impression that these men would never return,--the impression of all who witnessed their departure from the shore,--I watched their countenances closely. There was nothing approaching bravado in their looks, nothing to give a spectator any idea that they were about to engage in a matter of life or death, to themselves and the crew of the ship clinging to the fore-rigging of the *Northern Belle*. They had no hope of a decoration or of a pecuniary reward when, with a coolness of manner and a calmness of mind which contrasted strongly with the energy of their movements, they bounded into the lifeboat to storm batteries of billows far more appalling to the human mind than batteries surmounted by cannon and bristling with bayonets. There could be no question about the heroism of these men."

CHAPTER XVI.
A GALLANT RESCUE.

Shortly after daybreak, on the 4th January 1894, the lookout on the pier at Clacton-on-Sea saw a vessel strike on the Buxey Sand, about six miles from the shore. Without a moment's delay the warning was given, the lifeboat, *Albert Edward*, was manned and launched. There was need of the utmost speed. A strong easterly gale was raging at the time, accompanied by a nipping frost and blinding snowstorm. Owing to the extreme cold, it was feared that the shipwrecked crew would be unable to hold on till help arrived.

When the lifeboat reached the distressed vessel, it was found to be impossible to get alongside, so the coxswain ordered the anchor to be let go to windward. This was done, and the boat veered down to the full length of her cable. The waves continually broke over the vessel, and caused her to bump upon the sand in a frightful manner, thus preventing the lifeboat from approaching her. Under these circumstances, the boatmen decided to haul in the cable, and to drop the anchor nearer the vessel. This was a work of no little difficulty, and was rendered on this occasion highly dangerous by the anchor having fouled something on the sand. They tugged and strained for some time, but all to no purpose, and they were at last compelled to cut the rope. The sail was then set, and the lifeboat proceeded to the leeside of the ship.

There everything was in a terrible muddle, for the masts and rigging, which hung over the bulwarks, swayed about, threatening death to anyone who ventured within their reach. The sea was running too high to permit the men to board the ship, but by ebb-tide the coxswain thought that the sea would become smoother, and thus enable him to rescue the men at less risk. The crew of the vessel were nearly frozen to death, and it seemed as if they could not hold out much longer. The coxswain made signs to the poor fellows to fasten a buoy to a line, and slack it away from the ship towards the lifeboat. His signs were understood and promptly obeyed, but unfortunately the line caught in the rigging alongside and stuck fast.

The resources of the lifeboatmen were not yet exhausted. Sailing as close as possible to the vessel, they threw out a grappling line, which luckily caught on, and the boat was held. The coxswain shouted to the sailors to make another rope fast, but they paid no heed to his order. No sooner did they perceive that the boat was fixed than they began to crawl along the mast. Only one man had been taken on board, when a heavy sea swept down upon the lifeboat. The rope which fastened her to the wreck was not strong enough to bear the strain, and once more the *Albert Edward* was driven from the ship.

Canvas was again set to windward for about half an hour, and then the boat was headed for the wreck. The tide was now on the ebb, and less difficulty was experienced in getting a hold on the ship. One by one the poor fellows were taken on board the lifeboat, till only the captain remained. He was an old man, and so exhausted by suffering that he was unable to jump for the boat. A line was therefore thrown to him which he fastened round his waist, and the coxswain went to assist him over the rail of the ship. Just as he was in the act of performing this humane service he was knocked overboard by a sudden lurch. As he struggled in the water, he received a severe blow on the head and a wound across the eye from pieces of floating wreckage. His case was desperate, but he did not lose his presence of mind for a moment. Seizing hold of the rope which was made fast round the captain, he managed to keep himself afloat till his companions rescued him from his perilous position. Nothing daunted, he then made further efforts to save the captain, who was at length hauled through the surf and lifted on board in safety.

Just as this was accomplished, a heavy sea snapped the rope, and the lifeboat left the wreck, having on board the whole crew of seven men. In getting off the sands, on her homeward journey, the boat was frequently smothered by the heavy seas, and several of the men were badly hurt by being dashed against the side. At length, after a long, toilsome struggle, the harbour was reached, the lifeboat and her crew being covered with ice. In spite of the severity of the weather, a number of people were on the pier to give the heroes a hearty reception. The shipwrecked men, who were completely exhausted, were supplied with food and put to bed to recover from the effects of their exposure and fatigue. Their vessel was the St. Alexine of Copenhagen, bound for Stranraer with deals.

CHAPTER XVII.
A BUSY DAY.

In the early morning of the 7th of November 1890, while one of the severest storms known for years on the coast of Lancashire was at its height, signal flares were observed about three miles out at sea. A gun was fired to arouse the lifeboatmen, and in a few minutes the Fleetwood boat was launched and hurrying on her errand of mercy in the wake of a steam-tug. It was almost dark at the time, and the two vessels were quickly lost to view. The news rapidly spread that the lifeboat had been summoned, and soon a number of people were making their way to the beach in the hope of catching a sight of the distressed vessel.

It was not until seven o'clock that the hull of a large barque loomed in sight to those on shore, and it was then evident that but for the gallant services of the lifeboatmen all on board would be lost. Having got well to windward, the tow-rope was let go, and the boat drifted gradually down to the wreck. Here lay the real danger, and it required all the seamanship of the coxswain to prevent the boat from being dashed against the side of the ill-fated vessel, or swept past the mark by the force of the sea. When within a short distance, the boat was brought to an anchor, and veered down on her cable close to the wreck, which was found to be the *Labora*, a Norwegian ship.

The work of rescue was promptly begun, and as it was found to be utterly impossible for the lifeboat to approach near enough to take the men off, the coxswain shouted to the sailors to throw him a line. A lifebuoy was accordingly thrown overboard with a rope attached, and floated to the boat. Communication having been thus established, the crew were dragged through the surf in safety. The work of rescue lasted above two hours, and the boat was repeatedly filled with water, so that the fact that not a single life was lost reflects great credit on the seamanship of the coxswain and his men. The whole crew of the *Labora*, thirteen in number, were taken on board, the captain being the last man to leave the ship.

Sail was then hoisted on the lifeboat, and she made for the shore with all speed. Notwithstanding the gale and the driving rain, hundreds of spectators had assembled along the beach to await the return of the boat. When at length she appeared, she was greeted with shouts of joy, and landed the rescued crew amid a perfect salvo of cheering.

A few hours later, news of another wreck was brought to Fleetwood. Utterly regardless of their rough experience in the early morning, the crew again donned their lifebelts and manned the lifeboat. As they were towed out by the steamer, a magnificent sight was witnessed, the waves dashing furiously over the boat as she ploughed her way through the water, and both vessels were often completely hidden from sight by the seas breaking over them.

SIGHTING THE WRECK.

Regardless of the drenching they received, they held resolutely on their way, and soon the distance of five miles which intervened between them and the wreck was covered. The crew hailed the approach of the saving boat with loud cheers, but great difficulty was experienced in effecting the rescue, as all the masts and rigging were dashing about alongside the ship. To avoid the wreckage striking the lifeboat, and at the same time to get sufficiently near for the sailors to jump aboard, required great skill and judgment, as well as a cool head and a steady nerve.

Owing to the position in which the stranded vessel was lying, every sea broke over her, and threatened to swamp the lifeboat. Eventually the whole crew of eleven men were rescued, and the lifeboat was headed for the shore, where the crew were landed in a most exhausted condition. But for the brave efforts and untiring exertions of the lifeboatmen, the crews of both of those vessels would have been lost, and well might the noble fellows congratulate themselves on having within a few short hours saved twenty-four of their fellow-men from death.

CHAPTER XVIII.
A RESCUE IN MID-OCEAN.

It is a common belief at the present day that our sailors are no longer the same bold, kind-hearted fellows that they were before the introduction of steam and other modern improvements. From time to time, however, a brief account of some splendid act of heroic daring, performed on the high seas, finds its way into the newspapers, and proves that, after all, Jack is of the same race as the men who, in bygone days, won for England the proud title of "Mistress of the Seas."

Recently, while the Cunard steamer *Parthia* was crossing the Atlantic from America to England, her passengers had an opportunity of witnessing a genuine feat of derring-do of the old heroic kind. It was a Sunday afternoon, and for some hours the barometer had been steadily falling, a sure sign of a coming gale. Overhead the blue sky was dotted with white clouds, but away to the south and west the heavens were of a dull leaden colour.

About four o'clock, true to the indications it had given, the storm burst. The fury of the wind raised a tremendous sea, and after running for a time, it was judged prudent to bring the *Parthia* head on to the waves. All the passengers were ordered below lest they should be washed overboard, and the hatches were securely battened down to prevent the cabins being flooded. Every now and again the crew on deck were waist deep in water, as the steamer dipped her bows into the sea and took great surging waves on board.

For six hours the vessel lay-to, and during all that time the tempest raged with undiminished fury. The wind screamed and whistled mournfully through the rigging, and the mountainous waves dashed themselves with tremendous force against the sides of the ship, throwing the spray as high as the masthead At ten o'clock the gale moderated, and the steamer once more resumed her voyage. The night passed without further incident, and when the sun rose next morning out of the heaving waters it gave promise of a fair day.

Meanwhile a far different scene was being enacted on the angry ocean some miles away. A sailing ship was being tossed about like a plaything. One by one her sails were blown to ribbons, her planks sprung a-leak under the continued pounding of the waves, and as the vessel slowly settled down the crew gave themselves up for lost. As the water-logged hull tumbled about in the trough of the sea, they expected that she would go down every moment, but day broke and found them still afloat, looking for help in every direction and finding none. Assistance was, however, at hand.

All this time the *Parthia* had been steadily steaming on her homeward voyage. About nine o'clock in the morning the look-out man reported that a vessel was in sight. As the steamer approached, it became apparent to all on board that the ship was in distress. She lay low in the water, her rigging was all in a tangle, and upon the deck twenty-two wretched, pale-faced men could be counted, watching the steamer with wistful gaze. All these had to be saved, and every man on board the *Parthia* knew that this could only be done at the risk of the lives of those who went to their assistance, for a heavy sea was still running.

Few things are more perilous and difficult than lowering a boat during a storm in mid-ocean. The most seamen-like smartness may fail to save the frail fabric from being dashed to pieces against the iron side of the vessel, and even if the boat succeeds in getting away, the utmost skill is necessary to prevent her from being upset. Everyone of the *Parthia's* crew knew the danger, but not one of them shrank from the duty which faced them.

"Volunteers for the wreck!" shouted the captain, and in response to his summons eight men sprang forward and scrambled into the lifeboat. The third officer stepped into the stern, and took the rudder lines in his hands. Every man sat silent and ready while the boat swung from the davits. Calmly the order was given to lower, and the boat sank swiftly down to the water. As she rose on the crest of the next wave, the blocks were unhooked, and in another moment she was making for the wreck.

The passengers who thronged the deck of the *Parthia* watched the lifeboat in an agony of excitement. Now she disappeared as completely as if she had gone to the bottom; then she rose on the crest of a mighty billow, where she poised for an instant before taking the headlong plunge into the watery abyss beyond. A short struggle brought the boat within reach of the doomed vessel, and the mate shouted to the crew to heave him a line. It was caught, a lifebuoy was attached to it, and it was hauled on board the wreck. To the lifebuoy was tied a second line, one end of which was held by the lifeboat crew. The meaning of these arrangements soon became apparent. One of the shipwrecked sailors slipped his shoulders through the lifebuoy, plunged into the sea, and was dragged into the lifeboat. One by one the sailors were hauled on board, till eleven had been rescued. Then, with a cheering shout to those who were left behind, the boat returned to the steamer.

Meanwhile the captain of the *Parthia* had been busy making all the necessary preparations for taking the shipwrecked men on board. A rope with a loop at the end was suspended from the foreyard arm, and under this the lifeboat was stationed. The rope was then passed down, and the loop slipped under the arms of one of the men, who was then hoisted on board by the sailors.

When the first boatload had been safely deposited on the deck of the steamer, the lifeboat returned to the wreck. By means of the lifebuoys and lines the remainder of the crew were taken off, and afterwards hoisted on board the steamer in the same way as their companions. Her work having been accomplished, the lifeboat was hauled in, and the *Parthia* went "full speed ahead," to make up for lost time.

An eye-witness of this perilous and gallant rescue says:--

"To appreciate the pathos and pluck of an adventure of this kind, one must have served as a spectator or actor in some such scene. The expression on the faces of those shipwrecked men, as they were hoisted one by one over the *Parthia's* side; the bewildered rolling of their eyes, their expression of suffering, slowly yielding to the perception of the new lease of life mercifully accorded them, graciously and nobly earned for them; their streaming garments, their hair clotted like seaweed on their foreheads; the passionate pressing forward of the crew and passengers to rejoice with the poor fellows on their salvation from one of the most lamentable dooms to which the sea can sentence, will ever be vividly imprinted on the minds of those who witnessed the occurrence."

CHAPTER XIX.
THE "THREE BELLS."

Captain Leighton, of the British ship *Three Bells*, some years ago rescued the crew of an American vessel sinking in mid-ocean. Unable to take them off in the storm and darkness, he kept by them until morning, running down often during the night, as near to them as he dared, and shouting to them through his trumpet, "Never fear! hold on! I'll stand by you!"

Beneath the low-hung night-cloud
That raked her splintering mast,
The good ship settled slowly,
The cruel leak gained fast.

Over the awful ocean
Her signal guns pealed out.
Dear God! was that Thy answer
From the horror round about?

A voice came down the wild wind,
"Ho! ship ahoy!" its cry:
"Our stout *Three Bells* of Glasgow
Shall stand till daylight by!"

Hour after hour crept slowly,
Yet on the heaving swells
Tossed up and down the ship-lights,
The lights of the *Three Bells*.

And ship to ship made signals,
Man answered back to man,
While oft to cheer and hearten

The *Three Bells* nearer ran.

And the captain from her taffrail
Sent down his hopeful cry,
"Take heart! hold on!" he shouted,
"The *Three Bells* shall stand by!"

All night across the water
The tossing lights shone clear;
All night from reeling taffrail
The *Three Bells* sent her cheer.

And when the dreary watches
Of storm and darkness passed,
Just as the wreck lurched under,
All souls were saved at last.

Sail on, *Three Bells*, for ever,
In grateful memory sail!
Ring on, *Three Bells* of rescue,
Above the wave and gale!

J. G. WHITTIER.

CHAPTER XX.
ON THE CORNISH COAST.

One stormy December day, a few years ago, a horse reeking with foam galloped into Penzance, bearing a messenger with news that a ship which had got into the bay was unable to make her way out, and would in all probability be wrecked. The news spread through the quaint old town like wildfire, and in a few minutes hundreds of people were on the shore anxiously watching for the ship. From time to time she could be seen through the mist, and it was evident that her captain and crew were making every effort to head her out to the open sea; but there was little chance of success with such a furious gale blowing directly inshore. Anchors were thrown out in the hope of averting the threatened disaster, but they were of no use, and soon the vessel was drifting helplessly to the shore. "Man the lifeboat! man the lifeboat!" was then the cry, and coastguards and fishermen rushed off to the boathouse at full speed.

LIVES IN PERIL.

There was not a moment to spare. Horses were brought out and harnessed to the carriage, the men took their places, and away went the horses at full speed. The boat was launched into the breakers with a hearty cheer, and headed straight for the wreck.

Meanwhile a terrible tragedy was being enacted between the wreck and the shore, some distance to the east. The captain had seen two shore boats put off to his assistance, and after battling bravely with the sea for some time give up the attempt. He did not see the lifeboat, and, thinking that the safety of himself and his crew depended on their own efforts, he ordered one of the ship's boats to be lowered. No sooner had it touched the water than it was dashed to pieces against the side of the ship. A second boat was got out of the davits, and the captain and nine men got into her in safety, and made for the shore. She had not gone far when a huge wave pounced down upon her, whirled her round, and in another moment the men were struggling in the water, about three hundred yards from the shore. A few sailors seized the keel of the upturned boat, but again and again they were dashed from their hold by the heavy breakers, others seized the oars, and the captain struck out for the shore, followed by a few of his men. On the beach the people were helpless; but, seeing the captain swimming towards them, some of the strongest men joined hands, and waded out into the sea to meet him. One brave man, famous for miles round on account of his great strength, threw off his coat, and, followed by several others, dashed into the surf, determined to rescue at least one of the perishing sailors. When he got hold of one man he handed him over to his companions to be taken ashore, and, in defiance of the enormous breakers, he stayed out until he had rescued three men from certain death. Nine men reached the shore, but only four of those, who, full of health and strength, had put off from the wreck half an hour before, survived.

Now let us return to the lifeboat. "After a pull of more than an hour she reached the vessel. As she was pulling under her stern, a great sea struck the boat, and immediately capsized her. All on board were at once thrown out; the noble boat, however, at once self-righted. The coxswain was jammed under the boat by some wreckage, and very nearly lost his life, having to dive three or four times before he could extricate himself. When dragged on board, he was apparently dead, and in this state was brought ashore. Another man, pulling the stroke oar, was lost altogether from the boat, and the men were all so exhausted that they could not pull up to rescue him; but his cork jacket floated him ashore, when a brave man, named Desreaux, swam his horse out through the surf and rescued him.

"The inspecting-commander of the coastguard, who expressed an earnest wish to go off on this occasion, was also on board, and with others suffered severely. It is due to him to say that his great coolness and judgment, as well as his exertions, greatly aided in bringing the boat and her exhausted crew to shore. The second coxswain also behaved like a hero, and, though scarcely able to stand, managed the boat with the greatest skill when the coxswain was disabled.

"Judge of the dismay of those on shore when they saw the boat returning without having effected a rescue. It was at once clear that some disaster had happened, and they rushed to meet her. There was the coxswain, apparently dead, a stream of blood trickling from a wound in his temple, one man missing, and all the crew more or less disabled. Volunteers were at once called for. The second coxswain pluckily offered to go again, but this was not allowed, and his place was taken by the chief officer of the coastguard. In a short time another crew was formed, and the boat put off.

"No words can describe the struggle which followed. The boat had to be pulled to windward in the teeth of a tremendous gale. Sometimes she would rise almost perpendicular to the waves, and the people on shore looked on with bated breath, fearing she must go over. The way was disputed inch by inch, and at last the victory was won. Long and loud rang the cheers as the boat neared the shore, and quickly the shipwrecked mariners and their brave rescuers were safe.

"It was afterwards found that one of the second crew had three ribs broken, and several of the others had wounds and bruises more or less severe. Happily, none of the injuries proved fatal, and before long all the men, even the coxswain, went about their work as usual. The wrecked vessel was the *North Britain*, with a cargo of timber on board from Quebec."

CHAPTER XXI.
A PLUCKY CAPTAIN.

Lizard Point in Cornwall, the most southerly headland in England, is a piece of rocky land, which "has caused more vivid and varied emotions than any other on our coasts. The emigrant leaving, as he often thinks, his native land for ever; the soldier bound for distant battlefields, and the sailor for far-distant foreign ports, have each and all strained their eyes for a last parting glimpse of an isle they loved so much, and yet might never see again. And when the lighthouses' flash could no longer be discerned, how sadly did one and all turn into their berths to think--ay, 'perchance to dream'--of the happy past and the doubtful future.

"How different are the emotions of the homeward bound--the emigrant with his gathered gold, the bronzed veteran who has come out of the fiercest conflict unscathed, and the sailor who has safely passed the ordeal of fearful climes. The first glimpse of that strangely named rocky point is the signal for heartiest huzzas and congratulation."

There is, unfortunately, another side to this pleasant picture. Not unfrequently vessels become enveloped in the fogs, which prevail off this dangerous coast, and go crashing on to the rocks, there to become total wrecks. On the 4th of March 1893 an incident of this kind occurred. While the steamship *Gustav Bitter* of Newcastle-on-Tyne was proceeding from London to the Manchester Ship Canal with a general cargo, she stranded during a dense fog on the Callidges Rocks, off the Lizard Point. The engines were immediately reversed in the hope of getting her off, but she stuck fast. The captain gave the order for the long-boat to be lowered, and he got into her with seven men. As he was about to secure the boat's painter the rope was suddenly cut, and the strain being thus taken off, caused the captain to tumble into the sea, and he was compelled to swim to the boat to save his life. The second mate jumped from the deck of the doomed vessel, and tried to reach the boat, but unhappily he failed in the attempt, and was drowned.

News had already reached the shore that a ship was in danger, and the Polpear lifeboat was promptly manned and launched. When she reached the vessel the fog had lifted, and it was found that her bow was under water, and four men were clinging to the rigging. Great difficulty was experienced in getting near the vessel, as the seas were breaking completely over her and over the lifeboat. The lifeboatmen, however, succeeded in getting their grapnel on board, and the boat was brought up alongside. Three of the crew, watching their opportunity, left the rigging and went hand over hand along the grappling line from the steamer to the lifeboat. The fourth man, who is said to have been disabled by rheumatism, was unable to move from the rigging. His case was indeed desperate, for it was impossible to take the boat to the side of the ship on which he was lashed, on account of the shallowness of the water. To add to the difficulty of the situation, one of the men who had been rescued was in a very exhausted condition, and it was feared that he would not live much longer. After a little delay the boatmen decided, as there was no immediate danger of the vessel breaking up, that they would make for the shore, land the three men, and then return for the sufferer. The grapnel was accordingly freed from the rigging, and they pulled for the shore with all speed where the poor fellows were landed and well cared for. The lifeboat then proceeded on her return journey to the steamer.

Meanwhile another lifeboat had put off from the shore. On her way to the scene of action she fell in with the long-boat in which the captain and seven men had left the wreck. The little vessel was nearly half full of water and in great danger of being swamped, so her occupants were taken on board the lifeboat. They then told their rescuers that they had left four of their companions on board the steamer. Though the men were greatly exhausted with the hard pull of three miles which they had already performed, they gave a hearty shout and again bent their backs to the oars, and the remaining distance of a mile to the wreck was soon covered.

They of course were surprised to see only one man in the rigging instead of the four they had expected to find. The reason of his being where he was having been explained by the captain, several lifeboatmen volunteered for the dangerous task of rescuing the unfortunate man. The coxswain, however, thought it best to accept the offer of the captain, who was well acquainted with the ship, and had already proved himself a good swimmer. Two grapnels were thrown into the rigging of the steamer, and the captain swung himself on board by means of one of the lines. He reached the rigging, took the man out, and fastened a running line to his waist. Then he made a signal, and the poor fellow was hauled on board the lifeboat.

COMING ASHORE--"ALL SAVED!"

The captain was now compelled to take to the rigging again to avoid being washed overboard by the heavy seas, which were breaking over the ship. Twice he attempted to get off, but he was driven back each time. Watching his opportunity he tried again, and without either lifebelt or line plunged into the sea and swam to the boat. The work of rescue being then accomplished, the boat returned to the shore.

The silver medal of the Institution, accompanied by a copy of the vote inscribed on vellum, was awarded to Captain David Graham Ball, the master of the vessel, in recognition of his gallant conduct.

CHAPTER XXII.
BY SHEER STRENGTH.

During the terrific storm which spread such destruction over a large area of the United Kingdom in October 1889, a vessel was seen to be labouring heavily, and showing signals of distress, some two or three miles off the coast of Merionethshire. As she was rapidly drifting towards a very dangerous reef of rocks, the Aberystwyth lifeboatmen were speedily summoned. The tide was low at the time, and great difficulty was experienced in getting the boat to the water's edge. Several times she stuck in the soft sand, and the united exertions of the lifeboatmen could not move her forward a single inch. Plenty of willing helpers, however, were at hand, and after much labour and loss of valuable time, the boat was at length pushed into the sea on her carriage, and the crew took their places.

To avoid being blown on the rocks the men found it was necessary to row out for a considerable distance. The oars were manned, and the helpers eagerly waited for the word of command from the coxswain to let her go. The order was given; but here a fresh obstacle presented itself. The waves were rolling inshore with such fury that the greatest exertions of the boatmen failed to get her off, and notwithstanding the fact that scores of men went into the water till the waves broke over their heads, a considerable time passed before the boat could be got clear of her carriage and set afloat. Then the crew began a struggle against wind and waves, the like of which had not been seen for nine years, when one of the boatmen lost his life through exposure.

The men tugged at the oars with all their might, and seemed to be gaining slowly; but after they had been rowing for an hour they found themselves just where they started. Great white seas broke over the boat, drenching the men to the skin, and carrying her back towards the shore. Again and again the struggle was renewed, and again and again the boat was carried back on the crests of the waves. Sometimes the boat would be thrown on end, in an almost perpendicular position, and then fall into the trough of the sea and disappear.

For two hours the struggle against the angry sea and the fierce wind was kept up. During that time six oars were broken, and several times the boat narrowly escaped being upset. Then three huge rollers came in quick succession and carried the boat into the comparatively smooth water near the pier. She was brought alongside the landing-stage, and more oars and five additional men were taken on board.

As soon as the extra men were put in their places, another attempt was made to get the boat out to sea. The wind still blew with unabated force, and sea after sea broke over the little vessel. Slowly but steadily she made headway, and though she was often lost to sight in the trough of the sea, or buried in spray, she at length gained a point where the coxswain thought it was safe to hoist the sail. This was done, and away sped the lifeboat after the retreating vessel.

On getting alongside it was found that she was an American ship, and though terribly battered she was still holding on to her anchors. Two of the lifeboatmen were put on board to assist in navigating her, and, at the request of the captain, the boat remained alongside for some time, in order to be in readiness to save the crew in the event of the cables parting. While she was in this position an immense wave dashed right into the lifeboat, and three of the crew were swept overboard. They were afterwards picked up in a very exhausted condition.

Seeing that their services were not now required, the lifeboatmen cast off from the wreck and made for home, which was reached shortly before midnight. Their undaunted spirit won for them the admiration of the thousands of spectators who had watched their battle with the storm, and the owners of the vessel, wishing to show their appreciation of the crew's services, sent the sum of £30 "to be divided among the men as some slight recognition of their gallant conduct."

CHAPTER XXIII.
WRECKED IN PORT.

The spacious harbour of Milford Haven, on the south-west of Pembrokeshire, the finest in the kingdom, and large enough to shelter the whole British fleet, was, a few months ago, the scene of a most gallant rescue by a crew of South Wales lifeboatmen. On the 30th of January 1894, the full rigged iron ship *Loch Shiel* of Glasgow was stranded on Thorn Island, at the entrance to the Haven. She was bound for Australia with a general cargo, and had on board thirty-three persons, seven of whom were passengers.

As soon as the vessel struck, the captain tried the pump, and found that there was a quantity of water in the hold, and that the ship was rapidly sinking by the stern. He at once ordered the boats to be lowered. Then a mattress was brought on deck, soaked with paraffin oil, and lighted as a signal of distress. The flare was seen by the coastguard at St. Anne's Head, several miles away, and they telegraphed the news of the disaster to the lifeboat station at Angle. Obedient to the summons, the lifeboat put off to the rescue. Meanwhile several of the shipwrecked men had been forced to take refuge in the mizzen rigging, and others had climbed over the jibboom and landed on the rocks.

Presently the lifeboat came dashing along in splendid style. On nearing the vessel the anchor was dropped, and the boat's bow brought close to the mizzen rigging, to which six men could be seen clinging. One of these was an invalid passenger, and great difficulty was experienced in getting him on board. More than once the men expected to see him lose his hold and fall into the sea, but he, fortunately, had sufficient strength to hold on till he reached the arms stretched out to save him. The remaining sufferers were then quickly taken out of the top, the anchor was hauled in, and the boat pulled round to the leeside of the island, to take off the remainder of the crew and passengers.

Mr. Mirehouse, the Honorary Secretary of the Angle Branch of the Royal National Lifeboat Institution, who had accompanied the boat, and Edward Ball and Thomas Rees, two of the crew, now landed. Taking with them a rope and a lantern, they crawled along the edge of the cliff until they arrived above the spot where the people had taken refuge. They then lowered the rope over the cliff, and, in spite of the darkness of the night and the fury of the storm, they hauled up the remainder of the crew and passengers of the *Loch Shiel*, one of whom, a lady, was in a very weak and exhausted condition. But the rescue was not yet completed. The return journey had yet to be made along the narrow and dangerous pathway, in some parts barely a foot wide. The difficulties of the passage were further increased by having to guide the rescued and exhausted persons. To the credit of Mr. Mirehouse and his two men, be it told, that after great exertions and several narrow escapes they succeeded in bringing all in safety to the place where the lifeboat was in waiting.

As a very heavy surf was running, it was decided that the boat should make two trips. Twenty persons were accordingly put on board and landed at Angle. Then she returned immediately to the island for the remainder. At half-past six on the following morning she completed her second journey, and the whole thirty-three men and women were again in safety on the mainland. Some of the rescued people were taken to the residence of Mr. Mirehouse, and were most kindly cared for by him and his family; others were taken charge of by other residents.

Some time afterwards the following letter was received by Mr. Mirehouse from the captain of the vessel:--

GLASGOW, 21*st February* 1894.

DEAR SIR,--You and your dear lady, and your household, and all the inhabitants of Angle, please accept my humble thanks for the great kindness you all did to me and to my crew and passengers on the 30th and 31st January 1894; firstly, in taking us from the wreck of the ship *Loch Shiel*, on Thorn Island, and then having us at your house and other houses in Angle for some considerable time, thirty-three people in all.--I am, dear sir,

THOMAS DA VIES,
Master of the ill-fated ship *Loch Shiel* of Glasgow.

A highly gratifying letter was also received by the Honorary Secretary from the owners of the vessel, conveying their thanks for the services rendered to the crew and passengers. The crew of the ship also wrote expressing their thanks to the lifeboatmen for saving their lives, and to those who afterwards supplied them with food and clothing.

The silver medal of the Royal National Lifeboat Institution was awarded to Mr. Mirehouse, Thomas Rees, and Edward Ball in recognition of the bravery displayed by them, in going to the edge of the cliffs and rescuing the remainder of the passengers and crew, and in afterwards conducting them to a place of safety.

*   *   *   *   *

[The Royal Lifeboat Institution, the story of whose noble work we have followed, is supported solely by voluntary contributions, and to our credit as a nation be it said, that this admirable Society has never appealed in vain for funds to carry on its work. To the usual sources of revenue--annual subscriptions, donations, and legacies--another has been recently added, known as "Lifeboat Saturday." Originated in Manchester in 1891 by Mr. C. W. Macara, it rapidly spread from place to place, till now nearly every important town, both maritime and inland, sets apart one Saturday in each year to collect funds for this purpose. A procession is organised and one or two fully manned lifeboats are hauled through the streets, and where there is water launched at a convenient place. The presence of the boats and their crews never fails to arouse the greatest enthusiasm. The object of this movement is to further increase the funds of the Institution, that they may be able not only to reward the crews, but also in the event of loss of life, or permanent injury to health, to compensate those and all dependent on them for support. I have just been informed by the Secretary of the Royal National Lifeboat Institution that already this year (August 1894) they have granted rewards for saving nearly 500 lives. The lifeboatmen are all volunteers, and, as we have seen, each time they go out on service they literally take their lives in their hands. As the President of the Board of Trade recently said: "I trust the time will never come when the English public will abdicate their duty and their highest privilege of supporting such a noble Institution."]

THE END

Milton Keynes UK
Ingram Content Group UK Ltd.
UKHW040839141024
449705UK00006B/379